An Introduction to Advanced Quantum Physics

An Introduction to Advanced Quantum Physics

Hans P. Paar
University of California San Diego, USA

WILEY
A John Wiley and Sons, Ltd., Publication

This edition first published 2010

Registered office
John Wiley & Sons Ltd, The Atrium, Southern Gate, Chichester, West Sussex, PO19 8SQ, United Kingdom

For details of our global editorial offices, for customer services and for information about how to apply for permission to reuse the copyright material in this book please see our website at www.wiley.com.

Library of Congress Cataloging-in-Publication Data
Paar, Hans P.
An introduction to advanced quantum physics / Hans P. Paar.
p. cm.
Includes bibliographical references and index.
ISBN 978-0-470-68676-8 (cloth) – ISBN 978-0-470-68675-1 (pbk.)
1. Quantum theory – Textbooks. I. Title.
QC174.12P33 2010
530.12 – dc22
2009054392

A catalogue record for this book is available from the British Library.

ISBN: H/bk 978-0-470-68676-8 P/bk 978-0-470-68675-1

Typeset in 10/12 Sabon by Laserwords Private Limited, Chennai, India

2 2011

Contents

Preface

Over the years, material that used to be taught in graduate school made its way into the undergraduate curriculum to create room in graduate courses for new material that must be included because of new developments in various fields of physics. Advanced Quantum Physics and its relativistic extension to Quantum Field Theory are a case in point. This book is intended to facilitate this process. It is written to support a second course in Quantum Physics and attempts to present the material in such a way that it is accessible to advanced undergraduates and starting graduate students in Physics or Electrical Engineering.

This book consists of two parts.

Part 1, comprising Chapters 1 through 5, contains the material for a second course in Quantum Physics. This is where concepts from classical mechanics, electricity and magnetism, statistical physics, and quantum physics are pulled together in a discussion of the interaction of radiation and matter, selection rules, symmetries and conservation laws, scattering, relativistic quantum physics, questions related to the validity of quantum physics, and more. This is material that is suitable to be taught as part of an undergraduate quantum physics course for physics and electrical engineering majors. Surprisingly, there is no undergraduate textbook that treats this material at the undergraduate level, although it is (or ought to be) taught at many institutions.

In Part 2, comprising Chapters 6 through 8, we present elementary Quantum Field Theory. That this material should be studied by undergraduates is controversial but I expect it will become accepted practice in the future. This material is intended for undergraduates that are interested in the topics discussed and need it, for example, in a course on elementary particle physics or condensed matter. Traditionally such a course is taught in the beginning of graduate school. When teaching particle physics to advanced undergraduate students I felt that the time was ripe for an elementary introduction to quantum field theory, concentrating on only those topics that have an application in particle physics at that level. I have also taught the material in Chapters 6 through 8 to undergraduates and have found that they had no problem in understanding the material and doing the homework.

It is hoped that the presentation of the material is such that any good undergraduate student in physics or electrical engineering can follow it, and that such a student will be motivated to continue the study of quantum field theory beyond its present scope. Additionally, beginning graduate students may also find it of use.

Please communicate suggestions, criticisms and errors to the author at hpaar@ucsd.edu.

Hans P. Paar
January 2010

UNITS AND METRIC

It is customary in advanced quantum physics to use natural units. These are cgs units with Planck's constant divided by 2π and the velocity of light set to unity. Thus we set $\hbar = c = 1$.

In natural units we have for example that the Bohr radius of the hydrogen atom is $1/\alpha m$ with $\alpha = e^2/(4\pi)$ the fine-structure constant and m the mass of the electron. Likewise we have that the energy of the hydrogen atom is $\alpha m/(2n^2)$ and the classical radius of the electron is α/m.

When results of calculations have to be compared with experiment, we must introduce powers of $\hbar$ and c. This can be done easily with dimensional analysis using, for example, that the product $\hbar c = 0.197$ GeVfm and $(\hbar c)^2 = 0.389\,\mathrm{GeV}^2\mathrm{mb}$.

In summations I use the Einstein convention that requires one to sum over repeated indices, 1 to 3 for Latin and 1 to 4 for Greek letter indices.

Three-vectors are written bold faced such as $\mathbf{x}$ for the coordinate vector. We use the 'East-coast metric', introduced by Minkovski and made popular by Pauli in special relativity. Its name is obviously US-centric. In it, four-vectors have an imaginary fourth component. The alternative is to use a metric tensor as in General Relativity. This is called the 'West-coast metric', made popular by the textbook on Field Theory by Bjorken and Drell, but is overkill for our purposes. Thus $x_\mu = (\mathbf{x}, it)$, $p_\mu = (\mathbf{p}, iE)$, and $k_\mu = (\mathbf{k}, i\omega)$ with $\mathbf{k}$ the wave vector and ω the angular frequency. A traveling wave can be written as $\exp ikx = \exp i(k_x x + k_y y + k_z z - \omega t)$. Squaring four-vectors, we get for example $p^2 = p_\mu p_\mu = \mathbf{p}^2 - E^2$ which is $-m^2$ (unfortunately negative) and $k^2 = 0$ for electromagnetic radiation. When integrating in four-dimensional space we use d^4x which one might think is equal to $d^3\mathbf{x}idt$ but this is not so. By d^4x we mean $d^3\mathbf{x}dt$ (the West-coast metric does not have this inconsistency, sorry). With this convention the gamma matrices in the Dirac equation are all Hermitian.

We also find that we can write the commutation relations of $\mathbf{p}$ and x on the one hand and E and t on the other in relativistically covariant form as $[p_\mu, x_\nu] = \delta_{\mu\nu}/i$ where $\delta_{\mu\nu}$ is the Kronecker delta. Thus it is seen that the (at first sight odd) difference in sign of the two original commutation relations is required by relativistic invariance.

The partial derivatives $\partial/\partial x_\mu$ will often be abbreviated to ∂_μ. For example, the Lorenz condition $\nabla\mathbf{A} + \partial\phi/\partial t = 0$ can be written as $\partial_\mu A_\mu = 0$, showing that the Lorenz condition is relativistically covariant. Furthermore we have that $\partial^2\phi = \partial_\mu\partial_\mu\phi = (\nabla^2 - \partial^2/\partial t^2)\phi$, an expression that is useful in writing down a Lorenz invariant wave equation for the function ϕ.

In some equations the notation $\overline{\text{h.c.}}$ appears. This differs from the usual h.c. in that $\overline{\text{h.c.}}$ preserves the order of operators to which it is applied. So the h.c. of AB is $B^\dagger A^\dagger$ while the $\overline{\text{h.c.}}$ of AB is $A^\dagger B^\dagger$.

We do not follow the convention of some textbooks in which e stands for the absolute value of the electron charge; we use $e = -1.602 \times 10^{-19}$ C.

Part I

Relativistic Quantum Physics

1

Electromagnetic Radiation and Matter

1.1 HAMILTONIAN AND VECTOR POTENTIAL

The classical Hamiltonian describing the interaction of a particle with mass m and charge e with an electromagnetic field with vector potential $\mathbf{A}$ and scalar potential ϕ is

$$H = \frac{1}{2m}(\mathbf{p} - e\mathbf{A})^2 + e\phi \tag{1.1}$$

where $\mathbf{p}$ is the momentum of the particle. For example, $\mathbf{A}$ could be the vector potential of an external magnetic field while ϕ could be the Coulomb potential due to the presence of another charge. We do *not* follow the convention where e stands for the absolute value of the electron charge; in our case $e = -1.602 \times 10^{-19}$ C for an electron. This Hamiltonian is derived by casting the Lorenz force in the Hamiltonian formalism. It can also be obtained from the Hamiltonian of a free particle

$$H = \frac{\mathbf{p}^2}{2m} \tag{1.2}$$

with the substitutions

$$\mathbf{p} \to \mathbf{p} - e\mathbf{A} \qquad H \to H - e\phi \tag{1.3}$$

This is called the 'Minimal Substitution'. The substitutions (1.3) can be written in covariant form as

$$p_\mu \to p_\mu - eA_\mu \tag{1.4}$$

An Introduction to Advanced Quantum Physics Hans P. Paar

with $p_\mu = (\mathbf{p}, iE)$ and $A_\mu = (\mathbf{A}, i\phi)$, $c = 1$. The word 'minimal' indicates that no additional terms that would in principle be allowed are included (experiment is the arbiter). An example would be a term that accounts for the intrinsic magnetic moment of the particle. As is known from the study of an atom in an external static magnetic field, the Hamiltonian Equation (1.1) already accounts for a magnetic moment associated with orbital angular momentum. To obtain the Hamiltonian of the quantized system we use as always the replacements

$$\mathbf{p} \to \frac{1}{i}\nabla \qquad H\,(\text{or}\,E) \to -\frac{1}{i}\frac{\partial}{\partial t} \tag{1.5}$$

or in covariant form

$$p_\mu \to \frac{1}{i}\frac{\partial}{\partial x_\mu} = \frac{1}{i}\partial_\mu \tag{1.6}$$

The requirement that the substitutions in Equation (1.5) be covariant, that is, that they can be written in the form of Equation (1.6), explains the minus sign in Equation (1.5). The substitutions are a manifestation of 'First Quantization' in which momenta and energies, and functions dependent upon these, become operators. They lead to the commutation relations

$$[p_\mu, x_\nu] = \frac{1}{i}\delta_{\mu\nu} \tag{1.7}$$

where $\delta_{\mu\nu}$ is the Kronecker delta.

The question arises as to how to quantize the electromagnetic field. We know from the hypothesis of Planck and its extension by Einstein in the treatment of black body radiation and the photo-electric effect respectively that electromagnetic energy is quantized with quanta equal $\hbar\omega$. The replacements Equation (1.6) are of no use for an explanation. We will address this issue fully in the next subsection when we introduce 'Second Quantization' in which the vector and scalar potentials and thus the electric and magnetic fields become operators. This is a prototype of relativistic quantum field theory.

In preparation for quantization of the electromagnetic field, we will briefly review the arguments that lead to the wave equation for $\mathbf{A}$ with $\phi = 0$ (the Coulomb Gauge). Recall from classical electromagnetism that when the vector and scalar potentials are transformed into new ones by the Gauge transformation

$$\mathbf{A} \to \mathbf{A}' = \mathbf{A} + \nabla\chi \qquad \phi \to \phi' = \phi - \frac{\partial\chi}{\partial t} \tag{1.8}$$

that the values of electric and magnetic fields **E** and **B** do not change. This is so because under the Gauge transformation in Equation (1.8) we have

$$\mathbf{E} = -\nabla\phi - \frac{\partial \mathbf{A}}{\partial t} \rightarrow \mathbf{E}' = -\nabla\phi' - \frac{\partial \mathbf{A}'}{\partial t} = \mathbf{E} \tag{1.9}$$

and

$$\mathbf{B} = \nabla \times \mathbf{A} \rightarrow \mathbf{B}' = \nabla \times \mathbf{A}' = \mathbf{A} \tag{1.10}$$

where we used in Equation (1.9) that the order of ∇ and $\partial/\partial t$ can be exchanged while we used in Equation (1.10) that $\nabla \times \nabla\chi = 0$ for all χ. The function χ is arbitrary (unconstrained). The minus sign in Equation (1.8) is necessary for **E** to remain unchanged under the Gauge transformation. The minus sign also follows if we require that the relations Equation (1.8) can be written in a covariant form

$$A_\mu \rightarrow A'_\mu = A_\mu + \partial_\mu \chi \tag{1.11}$$

Writing out the fourth component of $A_\mu = (\mathbf{A}, i\phi)$ in Equation (1.11), one finds the second relation of Equation (1.8). One can make a Gauge transformation to find new $\mathbf{A}'$ and ϕ' such that they satisfy the Lorenz condition

$$\nabla \cdot \mathbf{A}' + \frac{\partial \phi'}{\partial t} = \partial_\mu A'_\mu = 0 \tag{1.12}$$

as follows. If $\nabla \cdot \mathbf{A} + \partial\phi/\partial t = f(\mathbf{x}, t) \neq 0$ then the new $\mathbf{A}'$ and ϕ' will satisfy the Lorenz condition in Equation (1.12) if χ is required to satisfy the inhomogeneous wave equation

$$\nabla^2 \chi - \frac{\partial^2 \chi}{\partial t^2} = -f(\mathbf{x}, t) \tag{1.13}$$

The Lorenz condition is seen to be covariant as well. The Lorenz condition simplifies the differential equations for the vector and scalar potentials to

$$\nabla^2 \mathbf{A} - \frac{\partial^2 \mathbf{A}}{\partial t^2} = -4\pi \mathbf{j} \qquad \nabla^2 \phi - \frac{\partial^2 \phi}{\partial t^2} = -4\pi\rho \tag{1.14}$$

or

$$\partial_\mu \partial_\mu A_\alpha = \partial^2 A_\alpha = 0 \tag{1.15}$$

showing that the wave equations for the potentials are covariant as they should be. This is the Lorenz Gauge.

There is more freedom left in the choice of **A** and ϕ in that a further Gauge transformation as in Equation (1.11) can be made that results in the scalar potential ϕ being zero if we require that the new χ satisfies

$$\phi = \frac{\partial \chi}{\partial t} \quad \text{and} \quad \nabla^2 \chi - \frac{\partial^2 \chi}{\partial t^2} = 0 \tag{1.16}$$

compared with Equation (1.13). This is the Coulomb or Radiation Gauge. The electric and magnetic fields are now given by

$$\mathbf{E} = -\frac{\partial \mathbf{A}}{\partial t} \qquad \mathbf{B} = \nabla \times \mathbf{A} \tag{1.17}$$

compared with the relations for **E** and **B** used in Equation (1.9) and Equation (1.10). The Lorenz condition in Equation (1.12) in the Coulomb Gauge is

$$\nabla \cdot \mathbf{A} = 0 \tag{1.18}$$

We introduce the wave vector **k** and the corresponding angular frequency ω with $\mathrm{k}^2 = \omega^2$. It follows from Equation (1.18) for a traveling wave of the form $\mathbf{A}(\mathbf{x}, t) = \mathbf{A}_0 \exp(i\mathbf{k} \cdot \mathbf{x} - \omega t)$ that

$$\mathbf{k} \cdot \mathbf{A} = 0 \tag{1.19}$$

so **k** and **A** are perpendicular to each other. From the first relation in Equation (1.17) we find that $\mathbf{E} = i\omega\mathbf{A}$, so **E** and **A** are parallel and thus **k** and **E** are also perpendicular. The second relation in Equation (1.17) gives $\mathbf{B} = i\mathbf{k} \times \mathbf{A} = \mathbf{k} \times \mathbf{E}/\omega$. This relation shows that **B** is in phase with **E** and is perpendicular to both **k** and **E**, so all three vectors **E**, **B**, **k** are mutually perpendicular. They form a right-handed triplet in that order because $\mathbf{E} \times \mathbf{B} = \mathbf{E} \times (\mathbf{k} \times \mathbf{E})/\omega = \mathrm{E}^2\mathbf{k}/\omega$ where we used that $\mathbf{E} \cdot \mathbf{k} = 0$. We define the polarization of the electromagnetic field as the direction of the electric field. This is so because the effects of the electric field dominate those of the magnetic field, for example in the exposure of photographic film. Because its electric field is perpendicular to its momentum we say that the electromagnetic field is transversely polarized. The cross product $\mathbf{E} \times \mathbf{B}$ equals the Poynting vector **S**, which is in the direction of **k** as it should be. Because Gauge transformations do not change the physical properties of the electromagnetic field, the last conclusion about the orientation of **E**, **B** and **k** holds as well in the Lorenz Gauge and indeed in general. Note that by transforming away ϕ we have not removed a Coulomb potential that might be present, we only removed the scalar potential associated with the electromagnetic field described by the coupled **E** and **B**.

The reader is urged to review this material from the text used in classical electromagnetism. A problem about Gauge transformations and the Coulomb Gauge is provided at the end of this chapter.

We stress that the sequence of two Lorenz transformations has left the electric and magnetic fields unchanged, so the physical properties of the system have not been affected. This is in analogy with elementary classical mechanics where it is shown that the potential energy $U(\mathbf{r})$ is defined up to a constant because adding a constant to U does not change the force $\mathbf{F} = -\nabla U$ and does not change the physical properties of the system.

The total Hamiltonian of the system consisting of a charged particle in an electromagnetic field with no other charges and currents but the ones associated with the particle in Equation (1.1) consists of the part given in Equation (1.1) and the energy of the electromagnetic field

$$E_{\mathrm{em}} = \frac{1}{8\pi} \int d^3\mathbf{x}(|\mathbf{E}|^2 + |\mathbf{B}|^2) \tag{1.20}$$

Because $\mathbf{E}$ or $\mathbf{B}$ may be complex we use absolute values in the integrand. The homogeneous wave Equation (1.14) for $\mathbf{A}$ becomes

$$\nabla^2\mathbf{A} - \frac{\partial^2\mathbf{A}}{\partial t^2} = 0 \tag{1.21}$$

This equation describes 'free' electromagnetic fields, that is, fields in the absence of currents and charges. We now seek solutions of the homogeneous wave Equation (1.21). We introduce the four-vector $k_\mu = (\mathbf{k}, i\omega)$ where $\mathbf{k}$ is the wave vector and ω the corresponding angular frequency. With $x_\mu = (\mathbf{x}, ict)$ we find that $kx = \mathbf{k}\cdot\mathbf{x} - \omega t$. Therefore we can write a traveling wave $\mathbf{A}(\mathbf{x}, t) = \mathbf{A}_0\exp(i\mathbf{k}\cdot\mathbf{x} - \omega t)$ as $\mathbf{A}(x) = \mathbf{A}_0\exp(ikx)$. Traveling waves with the vector potential of this form satisfy the wave equation provided that $\mathbf{k}^2 - \omega^2 = k^2 = 0$. This condition can be enforced in the solution by including a factor $\delta(k^2)$ where δ is the Dirac δ function. The solution must represent a three-dimensional vector so we make use of three mutually perpendicular unit vectors $\boldsymbol{\varepsilon}_\lambda(\mathbf{k})$ $(\lambda = 1, 2, 3)$. The Lorenz condition in the Coulomb Gauge Equation (1.19) requires that $\mathbf{k}$ and $\mathbf{A}$ are perpendicular to each other. If we choose $\boldsymbol{\varepsilon}_1$ and $\boldsymbol{\varepsilon}_2$ perpendicular to $\mathbf{k}$ (and to each other) and $\boldsymbol{\varepsilon}_3$ parallel to $\mathbf{k}$, such that $\boldsymbol{\varepsilon}_1, \boldsymbol{\varepsilon}_2, \boldsymbol{\varepsilon}_3$ form a right-handed set of unit vectors, terms proportional to $\boldsymbol{\varepsilon}_3$ must be absent because of Equation (1.19). Because the wave Equation (1.21) is linear, its most general solution is a linear superposition of terms of the form $a_\lambda(\mathbf{k}, \omega)\exp[i(\mathbf{kx} - \omega t)]\varepsilon_\lambda(\mathbf{k})\delta(k^2)$ $\lambda = 1, 2$. Our notation shows that the pre-factor $a_\lambda(\mathbf{k}, \omega)$ depends upon $\mathbf{k}$ and ω and that the unit vectors $\varepsilon_\lambda(\mathbf{k})$ depend upon (the direction of) $\mathbf{k}$. The most general solution can thus be written as

$$\mathbf{A}(\mathbf{x}, t) = \left(\frac{1}{2\pi}\right)^2 \int d^3\mathbf{k}\, d\omega \sum_{\lambda=1}^{2} a_\lambda(\mathbf{k})\, e^{i(\mathbf{k}\cdot\mathbf{x} - \omega t)} \boldsymbol{\varepsilon}_\lambda(\mathbf{k})\, \delta(k^2) \tag{1.22}$$

The $\boldsymbol{\varepsilon}_\lambda(\mathbf{k})$ are called polarization vectors. One can interpret this solution as a four-dimensional version of the Fourier transform which is familiar from the solution of a wave equation in one dimension. The one-dimensional Fourier transform has a pre-factor $1/\sqrt{2\pi}$, hence the $(1/2\pi)^2$ in Equation (1.22). This is the integral version of the Fourier transform. Substitution of Equation (1.22) in the wave Equation (1.21) shows that Equation (1.22) is indeed a solution if the condition $k^2\delta(k^2) = 0$ is satisfied for each term separately. This condition is satisfied for the Dirac-δ function as in general $x\,\delta(x) = 0$ because the Dirac-$\delta(x)$ function is even in x and it is multiplied by the odd function x. The general solution in Equation (1.22) satisfies the condition in Equation (1.19) by construction. We know that **A** and **E** are parallel, see Equation (1.17), so the direction of **A** as specified by $a_1(\mathbf{k},\omega)\boldsymbol{\varepsilon}_1(\mathbf{k})$ and $a_2(\mathbf{k},\omega)\boldsymbol{\varepsilon}_2(\mathbf{k})$ is the direction of polarization.

We can simplify Equation (1.22) by integrating over ω using the constraint provided by the factor $\delta(k^2) = 0$. The condition $k^2 = \mathbf{k}^2 - \omega^2 = 0$ leads to the requirement that $\omega = \pm|\mathbf{k}|$. This is also expressed by the property of the Dirac-δ function

$$\delta(k^2) = \delta(\mathbf{k}^2 - \omega^2) = \frac{1}{2\omega_k}\big[\delta(|\mathbf{k}| + \omega) + \delta(|\mathbf{k}| - \omega)\big] \tag{1.23}$$

We introduce $\omega_k = |\mathbf{k}|$ and obtain

$$\begin{aligned}\mathbf{A}(\mathbf{x},t) &\\ &= \left(\frac{1}{2\pi}\right)^2 \int \frac{d^3\mathbf{k}d\omega}{2\omega_k} \sum_{\lambda=1}^{2} a_\lambda(\mathbf{k},\omega)\, e^{i(\mathbf{k}\cdot\mathbf{x}-\omega t)}\big[\delta(\omega_k+\omega) + \delta(\omega_k-\omega)\big]\boldsymbol{\varepsilon}_\lambda(\mathbf{k}) \\ &= \left(\frac{1}{2\pi}\right)^2 \int \frac{d^3\mathbf{k}}{2\omega_k} \sum_{\lambda=1}^{2}\big[a_\lambda(\mathbf{k},-\omega_k)\, e^{i(\mathbf{k}\cdot\mathbf{x}+\omega_k t)} + a_\lambda(\mathbf{k},\omega_k)\, e^{i(\mathbf{k}\cdot\mathbf{x}-\omega_k t)}\big]\boldsymbol{\varepsilon}_\lambda(\mathbf{k})\end{aligned} \tag{1.24}$$

We call the first term in Equation (1.24) with the negative value of ω the negative energy solution, while we call the second term in Equation (1.24) with the positive value of ω the positive energy solution. It is conventional to remove a factor $1/\sqrt{2\pi}$ and a factor $1/\sqrt{2\omega_k}$ and to add a factor $\sqrt{4\pi}$ in Equation (1.24). This merely redefines the coefficients a_λ. The reason for the first change is that the Fourier integral over $d^3\mathbf{k}$ ought to be accompanied by a pre-factor $1/\sqrt{2\pi}$ for each integration variable. The reason for the other two changes will become clear in the next subsection.

We would like to reinstate the compact notation $\exp(ikx)$ in Equation (1.24). The positive energy term is already of this form but the negative energy term is not because the terms $\mathbf{k}\cdot\mathbf{x}$ and $\omega_k t$ in the exponential do not have opposite signs. We can arrange for that by introducing a new integration variable $\mathbf{k}' = -\mathbf{k}$ in the negative energy term only. This gives three minus signs from $d^3\mathbf{k}' = -d^3\mathbf{k}$ and changes the signs of the integration limits in each of the three integrals. Exchanging the new upper and lower integration limits in each of the three integrals gives another three minus signs. The net result is that all minus signs cancel. We drop the prime in $\mathbf{k}'$. The argument of a_λ in the negative energy term will now be $-\mathbf{k}$. Another way to see this is to consider that we are looking for solutions of the wave equation in a large box with periodic boundary conditions at its surface. Solutions with $+\mathbf{k}$ and $-\mathbf{k}$ have the same physical properties.

The expression for $\mathbf{A}$ becomes

$$\mathbf{A}(x) = \left(\frac{1}{2\pi}\right)^{\frac{3}{2}} \sqrt{4\pi} \int \frac{d^3\mathbf{k}}{\sqrt{2\omega_k}} \sum_{\lambda=1}^{2} \left[a_\lambda(-\mathbf{k}, -\omega_k)\, \mathrm{e}^{-ikx} \boldsymbol{\varepsilon}_\lambda(-\mathbf{k}) + a_\lambda(\mathbf{k}, \omega_k)\, \mathrm{e}^{ikx} \boldsymbol{\varepsilon}_\lambda(\mathbf{k})\right] \tag{1.25}$$

$\mathbf{A}$ is used to calculate $\mathbf{E}$ and $\mathbf{B}$ according to Equation (1.17). In the next section $\mathbf{A}$, $\mathbf{E}$ and $\mathbf{B}$ become operators. The eigenvalues of $\mathbf{E}$ and $\mathbf{B}$ are observable, so the operators $\mathbf{E}$ and $\mathbf{B}$ and therefore $\mathbf{A}$ must be Hermitian operators. This means that $\mathbf{A}$ must be real quantity and we require that $\mathbf{A} = \mathbf{A}^*$. We find that

$$\mathbf{A}^*(x) = \left(\frac{1}{2\pi}\right)^{\frac{3}{2}} \sqrt{4\pi} \int \frac{d^3\mathbf{k}}{\sqrt{2\omega_k}} \sum_{\lambda=1}^{2} \left[a^*_\lambda(-\mathbf{k}, -\omega_k)\, \mathrm{e}^{ikx} \boldsymbol{\varepsilon}^*_\lambda(-\mathbf{k}) + a^*_\lambda(\mathbf{k}, \omega_k)\, \mathrm{e}^{-ikx} \boldsymbol{\varepsilon}^*_\lambda(\mathbf{k})\right] \tag{1.26}$$

Comparing Equation (1.26) with Equation (1.25) and using that the exponentials $\exp(\pm ikx)$ are orthogonal to each other, we find that

$$\sum_\lambda a_\lambda(-\mathbf{k}, -\omega_k)\boldsymbol{\varepsilon}_\lambda(-\mathbf{k}) = \sum_\lambda a^*_\lambda(\mathbf{k}, \omega_k)\boldsymbol{\varepsilon}^*_\lambda(\mathbf{k}) \tag{1.27}$$

$$\sum_\lambda a_\lambda(\mathbf{k}, \omega_k)\boldsymbol{\varepsilon}_\lambda(\mathbf{k}) = \sum_\lambda a^*_\lambda(-\mathbf{k}, -\omega_k)\boldsymbol{\varepsilon}^*_\lambda(-\mathbf{k}) \tag{1.28}$$

The two relations are each other's complex conjugate, so there is just one condition on a_λ and its complex conjugate. We will not attempt to formulate relations between individual terms in the sums over λ because we do not need those, and if we try it would lead to a left-handed set of three unit vectors $\boldsymbol{\varepsilon}$ or the appearance of minus signs in nasty places. Using Equation (1.27) in Equation (1.25) we have the result

$$\mathbf{A}(x) = \left(\frac{1}{2\pi}\right)^{\frac{3}{2}} \sqrt{4\pi} \int \frac{d^3\mathbf{k}}{\sqrt{2\omega_k}} \sum_{\lambda=1}^{2} \left[a_\lambda(\mathbf{k})\, \mathrm{e}^{ikx} \boldsymbol{\varepsilon}_\lambda(\mathbf{k}) + a^*_\lambda(\mathbf{k})\, \mathrm{e}^{-ikx} \boldsymbol{\varepsilon}^*_\lambda(\mathbf{k})\right] \tag{1.29}$$

We shall also use the form where a discrete sum over **k** with an infinite number of terms is used. This corresponds to quantization in a box with length L. The prefactor in a one-dimensional Fourier series becomes $1/\sqrt{L}$ instead of $1/\sqrt{2\pi}$ so in three dimensions we get a pre-factor $1/\sqrt{V}$. One can consider the integral form of the Fourier transform as the limiting case where L (and V) go to infinity and the sum becomes an integral.

If we use a Fourier series instead of a Fourier integral we get

$$\mathbf{A}(x) = \sqrt{\frac{4\pi}{V}} \sum_{\mathbf{k},\lambda} \frac{1}{\sqrt{2\omega_k}} [a_\lambda(\mathbf{k})\, \mathrm{e}^{ikx} \boldsymbol{\varepsilon}_\lambda(\mathbf{k}) + a_\lambda^*(\mathbf{k})\, \mathrm{e}^{-ikx} \boldsymbol{\varepsilon}_\lambda^*(\mathbf{k})] \tag{1.30}$$

The summation over λ is understood to be from 1 to 2. We do not explicitly show the dependence of a_λ (and a_λ^*) upon ω_k because $\omega_k = |\mathbf{k}|$ so it is specified by **k**. We allow for complex $\boldsymbol{\varepsilon}_\lambda(\mathbf{k})$ for applications with circular polarized electromagnetic radiation.

The $\boldsymbol{\varepsilon}_\lambda(\mathbf{k})$ are real for linearly polarized radiation, and can in that case be taken outside the square bracket part of Equation (1.29) and Equation (1.30). The expressions Equation (1.29) and Equation (1.30) explicitly show that **A** is real because **A** is the sum of two terms that are each other's complex conjugate. We will sometimes write 'h.c.' (Hermitian conjugate) for the second term.

We have discussed how to quantize the part of the classical Hamiltonian given in Equation (1.1). We will now quantize the classical Hamiltonian Equation (1.20) after which **A** and H_{em} are operators. We shall see that the operator **A** and thus H_{em} create and annihilate photons. Very interesting stuff indeed!

1.2 SECOND QUANTIZATION

1.2.1 Commutation Relations

Planck's 1901 treatment of black-body radiation and Einstein's 1905 discussion of the photo-electric effect assume that electromagnetic energy Equation (1.20) is quantized. Likewise, Compton scattering shows that the momentum of the electromagnetic field is quantized as well. Besides the fact that quantization itself needs to be 'explained', both quantizations have specific values for their quanta whose values need to be 'explained' as well. The quantizations of energy and momentum of the electromagnetic field and the numerical values of their quanta are introduced at the start of most Quantum Physics courses, but they are not derived there. This was also the case historically: Second Quantization, which provides this derivation, was not developed until about 1930. The quantization of the electromagnetic field contains many subtleties that have given rise to a variety of sophisticated treatments. We will not do that here but rather follow a more heuristic path.

We know from the treatment of the Harmonic Oscillator, using the operator formalism as opposed to the solutions using Hermite polynomials, that we can lower and raise the energy of the system with the operators A and $A^\dagger$, one quantum of energy $\hbar\omega$ at a time. This is the time to review that material in your favorite textbook. There appears to be some similarity with a quantized electromagnetic field whose energy can be increased or decreased by the creation or annihilation of photons. This similarity will be exploited. The A and $A^\dagger$ operators of the Harmonic Oscillator satisfy commutation relations that were derived from the commutation relations between momentum and position in Equation (1.5). In Second Quantization we postulate commutation relations between the coefficients a_λ and a_λ^* in Equation (1.29) and Equation (1.30) that are the same as those for A and $A^\dagger$. This means of course that the a and a^* become operators, so we write Hermitian conjugate $a^\dagger$ instead of complex conjugate a^*. Because the $a_\lambda(\mathbf{k})$ and $a_\lambda^\dagger(\mathbf{k})$ are operators, so is **A** and therefore are **E** and **B**! The formalism of second quantization starts with postulating commutation relations

$$[a_\lambda(\mathbf{k}), a_{\lambda'}^\dagger(\mathbf{k}')] = \delta_{\mathbf{k}\mathbf{k}'}\delta_{\lambda\lambda'} \tag{1.31}$$

$$[a_\lambda(\mathbf{k}), a_{\lambda'}(\mathbf{k}')] = [a_\lambda^\dagger(\mathbf{k}), a_{\lambda'}^\dagger(\mathbf{k}')] = 0 \tag{1.32}$$

The second relation in Equation (1.32) is the Hermitian conjugate of the first one in Equation (1.32), so it provides no additional information. These commutation relations involve momenta **k** and polarizations λ but are otherwise identical to those for the Harmonic Oscillator.

The same commutation relations lead to the same results so there is no need to derive these results again here. This is one reason why it is important to treat the Harmonic Oscillator with the operator formalism, even though analytic solutions of the Schödinger equation exist that have the same physical content. Thus we have the following relations

$$a_\lambda^\dagger(\mathbf{k})\,|n_\lambda(\mathbf{k})\rangle = \sqrt{n_\lambda(\mathbf{k})+1}\,|n_\lambda(\mathbf{k})+1\rangle \tag{1.33}$$

$$a_\lambda(\mathbf{k})\,|n_\lambda(\mathbf{k})\rangle = \sqrt{n_\lambda(\mathbf{k})}\,|n_\lambda(\mathbf{k})-1\rangle \tag{1.34}$$

where the kets form an orthonormal set. We call the $a^\dagger$ and a creation and annihilation operators instead of raising and lowering operators because they are thought to create and annihilate quanta (photons). Also in analogy with the Harmonic Oscillator, we have an operator N defined by

$$N_\lambda(\mathbf{k}) = a_\lambda^\dagger(\mathbf{k})\,a_\lambda(\mathbf{k}) \tag{1.35}$$

called the number operator. It satisfies the commutation relations

$$[a_\lambda(\mathbf{k}), N_{\lambda'}(\mathbf{k}')] = \delta_{\mathbf{k}\mathbf{k}'}\delta_{\lambda\lambda'}\,a_\lambda(\mathbf{k}) \tag{1.36}$$

$$[a_\lambda^\dagger(\mathbf{k}), N_{\lambda'}(\mathbf{k}')] = -\delta_{\mathbf{k}\mathbf{k}'}\delta_{\lambda\lambda'}\,a_\lambda^\dagger(\mathbf{k}) \tag{1.37}$$

The eigenvalues $n_\lambda(\mathbf{k})$ of the number operator $N_\lambda(\mathbf{k})$ represent the number of photons with momentum $\mathbf{k}$ and polarization λ that are present. A single photon of momentum $\mathbf{k}$ and polarization λ is represented by

$$|\mathbf{k}, \lambda\rangle = a_\lambda^\dagger(\mathbf{k})|0\rangle \tag{1.38}$$

where $|0\rangle$ represents the vacuum, a state without any photons. A state containing two different photons is represented by

$$|\mathbf{k}_1, \lambda_1; \mathbf{k}_2, \lambda_2\rangle = a_{\lambda_1}^\dagger(\mathbf{k}_1)\, a_{\lambda_2}^\dagger(\mathbf{k}_2)\, |0\rangle \tag{1.39}$$

while a state containing two identical photons is represented by

$$|\mathbf{k}_1, \lambda_1; \mathbf{k}_2, \lambda_2\rangle = \frac{a_{\lambda_1}^\dagger(\mathbf{k}_1)\, a_{\lambda_2}^\dagger(\mathbf{k}_2)}{\sqrt{2}}\, |0\rangle \tag{1.40}$$

as shown in the study of identical particles, in this case bosons. In general, for a state containing $n_\lambda(\mathbf{k})$ identical photons, we get a factor $\sqrt{n_\lambda(\mathbf{k})!}$ in the denominator. The second commutation relation in Equation (1.32) shows that one may exchange the $\mathbf{k}, \lambda$ labels (equal or not) of any two photons in such an expression, indicating that the multi-photon state satisfies Bose-Einstein statistics as required. You might at this point suspect that creation and annihilation operators for Fermions should satisfy anticommutation relations. If we did not run out of time before we have a chance to 'Second Quantize' the Dirac equation, you would discover that your suspicion was correct. You would even discover that the number operator associated with such anticommuting operators can only take on two values: 0 or 1: the Pauli principle 'derived'. Replacing the complex conjugate a^* by $a^\dagger$ in Equation (1.30) we get

$$\mathbf{A}(\mathbf{x}, t) = \sqrt{\frac{4\pi}{V}} \sum_{\mathbf{k},\lambda} \frac{1}{\sqrt{2\omega_k}} \left[a_\lambda(\mathbf{k})\, \mathrm{e}^{ikx} \boldsymbol{\varepsilon}_\lambda(\mathbf{k}) + a_\lambda^\dagger(\mathbf{k})\, \mathrm{e}^{-ikx} \boldsymbol{\varepsilon}_\lambda^*(\mathbf{k})\right] \tag{1.41}$$

This expression for $\mathbf{A}(\mathbf{x}, t)$ is our final result and will be used whenever we discuss the quantized electromagnetic field.

1.2.2 Energy

We now derive the quantization of the energy and the value of its quanta. The energy of an electromagnetic field in a vacuum is given by the classical expression in Equation (1.20). The values for $\mathbf{E}$ and $\mathbf{B}$ needed are calculated from Equation (1.17) where we use Equation (1.41) for **A**. When taking derivatives of **A** we will use the relations

$$\frac{\partial}{\partial t} e^{\pm ikx} = \mp i\omega_k e^{\pm ikx} \quad (1.42)$$

$$\nabla \times [\mathrm{e}^{\pm ikx}\boldsymbol{\varepsilon}_\lambda(\mathbf{k})] = \pm i\mathbf{k} \times [\mathrm{e}^{\pm ikx}\boldsymbol{\varepsilon}_\lambda(\mathbf{k})] \quad (1.43)$$

Therefore

$$\mathbf{E} = i\sqrt{\frac{4\pi}{V}} \sum_{\mathbf{k},\lambda} \sqrt{\frac{\omega_k}{2}} \left[a_\lambda(\mathbf{k})\mathrm{e}^{ikx}\boldsymbol{\varepsilon}_\lambda(\mathbf{k}) - a^\dagger_\lambda(\mathbf{k})\mathrm{e}^{-ikx}\varepsilon^*_\lambda(\mathbf{k})\right] \quad (1.44)$$

$$\mathbf{B} = i\sqrt{\frac{4\pi}{V}} \sum_{\mathbf{k},\lambda} \sqrt{\frac{1}{2\omega_k}} \left[a_\lambda(\mathbf{k})\mathrm{e}^{ikx}\mathbf{k} \times \boldsymbol{\varepsilon}_\lambda(\mathbf{k}) - a^\dagger_\lambda(\mathbf{k})\mathrm{e}^{-ikx}\mathbf{k} \times \varepsilon^*_\lambda(\mathbf{k})\right] \quad (1.45)$$

Notice that the second terms in Equation (1.44) and Equation (1.45) are the Hermitian conjugates of their respective first terms, as they should be, because in the expression for **A** the same is true and **E** and **B** are derived from **A**. Classically the $\mathbf{E}^2$ and $\mathbf{B}^2$ in Equation (1.20) contribute equally to the energy. We might therefore expect that to be the case here too, so only one of two terms would need to be calculated. This expectation is not correct as will be shown below by calculating both terms. We begin with the simpler $|\mathbf{E}|^2$ term using Equation (1.44)

$$|\mathbf{E}|^2 = \mathbf{E} \cdot \mathbf{E}^\dagger = \tfrac{4\pi}{V} \textstyle\sum_{\mathbf{k},\lambda} \sqrt{\tfrac{\omega_k}{2}} \left[a_\lambda(\mathbf{k})\mathrm{e}^{ikx}\varepsilon_\lambda(\mathbf{k}) - a^\dagger_\lambda(\mathbf{k})\mathrm{e}^{-ikx}\boldsymbol{\varepsilon}^*_\lambda(\mathbf{k})\right] \cdot$$

$$\textstyle\sum_{\mathbf{k}',\lambda'} \sqrt{\tfrac{\omega'_k}{2}} \left[a^\dagger_{\lambda'}(\mathbf{k}')\mathrm{e}^{-ik'x}\boldsymbol{\varepsilon}^*_{\lambda'}(\mathbf{k}') - a_{\lambda'}(\mathbf{k}')\mathrm{e}^{ik'x}\boldsymbol{\varepsilon}_{\lambda'}(\mathbf{k}')\right] \quad (1.46)$$

When working out this expression we get four terms, each summed over $\mathbf{k}, \lambda, \mathbf{k}', \lambda'$. There are two types of exponentials among the four terms: those with the difference and those with the sum of k and k' in the exponent. We reorder Equation (1.46) with the terms involving the difference $k - k'$ first

$$|\mathbf{E}|^2 = \frac{4\pi}{2V} \sum_{\mathbf{k},\mathbf{k}'\lambda\lambda'} \sqrt{\omega_k\omega'_k} \left[a_\lambda(\mathbf{k})a^\dagger_{\lambda'}(\mathbf{k}')\mathrm{e}^{i(k-k')x}\boldsymbol{\varepsilon}_\lambda(\mathbf{k}) \cdot \boldsymbol{\varepsilon}^*_{\lambda'}(\mathbf{k}') + \overline{\text{h.c.}}\right]$$

$$-\frac{4\pi}{2V} \sum_{\mathbf{k},\mathbf{k}',\lambda,\lambda'} \sqrt{\omega_k\omega'_k} \left[a_\lambda(\mathbf{k})a_{\lambda'}(\mathbf{k}')\mathrm{e}^{i(k+k')x}\boldsymbol{\varepsilon}_\lambda(\mathbf{k}) \cdot \boldsymbol{\varepsilon}_{\lambda'}(\mathbf{k}') + \overline{\text{h.c.}}\right] \quad (1.47)$$

The symbol $\overline{\text{h.c.}}$ indicates the Hermitian conjugate of the previous term but with the proviso, indicated by the line over h.c., that in a product of operators their original order is preserved. Thus the h.c. of $a_\lambda(\mathbf{k})a^\dagger_{\lambda'}(\mathbf{k}')$ is

$a_{\lambda'}(\mathbf{k}')a^\dagger_\lambda(\mathbf{k})$ while the $\overline{\text{h.c.}}$ of $a_\lambda(\mathbf{k})a^\dagger_{\lambda'}(\mathbf{k}')$ is $a^\dagger_\lambda(\mathbf{k})a_{\lambda'}(\mathbf{k}')$. That these are not the same for all $\mathbf{k}, \mathbf{k}', \lambda, \lambda'$ will be obvious from the commutation relations Equation (1.31). We must integrate the four terms over all of space. We exchange the order of the integration and the summations so as to do the integrals first. We use that

$$\int d^3\mathbf{x}\, e^{i(\mathbf{k}\pm\mathbf{k}')\cdot\mathbf{x}} = V\delta_{\mathbf{k}\mathbf{k}'} \tag{1.48}$$

Thus the double sum over $\mathbf{k}$ and $\mathbf{k}'$ reduces to a single sum over $\mathbf{k}$ and we have two types of terms: those with $\mathbf{k}' = \mathbf{k}$ and those with $\mathbf{k}' = -\mathbf{k}$. In the former the $\exp(-i\omega_k t)$ and $\exp(+i\omega_k t)$ factors cancel, while in the latter we get factors $\exp(-2i\omega_k t)$ and $\exp(+2i\omega_k t)$. Note that ω_k is defined just below Equation (1.23) as $\omega_k = |\mathbf{k}|$ so ω_k is positive, irrespective of the sign of $\mathbf{k}$. We get

$$\begin{aligned}\int d^3\mathbf{x}\, |\mathbf{E}|^2 = 2\pi \sum_{\mathbf{k},\lambda,\lambda'} \omega_k \left[a_\lambda(\mathbf{k})a^\dagger_{\lambda'}(\mathbf{k})\,\{\boldsymbol{\varepsilon}_\lambda(\mathbf{k})\cdot\boldsymbol{\varepsilon}^*_{\lambda'}(\mathbf{k})\} + \overline{\text{h.c.}}\,\right] \\ -2\pi \sum_{\mathbf{k},\lambda} \omega_k \left[a_\lambda(\mathbf{k})a_{\lambda'}(-\mathbf{k})\,e^{-2i\omega_k t}\{\boldsymbol{\varepsilon}_\lambda(\mathbf{k})\cdot\boldsymbol{\varepsilon}_{\lambda'}(-\mathbf{k})\} + \overline{\text{h.c.}}\,\right]\end{aligned} \tag{1.49}$$

In the first sum the double sum over λ and λ' reduces to a single sum over λ because

$$\boldsymbol{\varepsilon}_\lambda(\mathbf{k})\cdot\boldsymbol{\varepsilon}^*_{\lambda'}(\mathbf{k}) = \delta_{\lambda\lambda'} \tag{1.50}$$

This relation holds for linear polarization as well as for circular and elliptical polarization. Note that one and only one factor in the dot product has a complex conjugation sign and that the two factors have the same argument ($\mathbf{k}$). The complex conjugation of one of the two $\boldsymbol{\varepsilon}$ in Equation (1.50) is necessary when $\boldsymbol{\varepsilon}$ is complex and it conforms with the definition of the dot product of complex vectors (compare the scalar product of two 'vectors' in Hilbert space). The second sum in Equation (1.49) will cancel a similar term in the expression for $|\mathbf{B}|^2$ so we do not have to deal with the dot products of two $\boldsymbol{\varepsilon}_\lambda(\mathbf{k})$ or two $\boldsymbol{\varepsilon}^*_\lambda(\mathbf{k})$ that also have different arguments ($\mathbf{k}$). Compare with the comments below Equation (1.27) and Equation (1.28).

We calculate the $|\mathbf{B}|^2 = \mathbf{B}\cdot\mathbf{B}^\dagger$ term using Equation (1.45)

$$\begin{aligned}|\mathbf{B}|^2 = \frac{4\pi}{V}\sum_{\mathbf{k},\lambda}\sqrt{\frac{1}{2\omega_k}}\left[a_\lambda(\mathbf{k})e^{ikx}\mathbf{k}\times\boldsymbol{\varepsilon}_\lambda(\mathbf{k}) - a^\dagger_\lambda(\mathbf{k})e^{-ikx}\mathbf{k}\times\boldsymbol{\varepsilon}^*_\lambda(\mathbf{k})\right]\cdot \\ \sum_{\mathbf{k}',\lambda'}\sqrt{\frac{1}{2\omega'_k}}\left[a^\dagger_{\lambda'}(\mathbf{k}')e^{-ik'x}\mathbf{k}'\times\boldsymbol{\varepsilon}^*_{\lambda'}(\mathbf{k}') - a_{\lambda'}(\mathbf{k}')e^{ik'x}\mathbf{k}'\times\boldsymbol{\varepsilon}_{\lambda'}(\mathbf{k}')\right]\end{aligned} \tag{1.51}$$

When working out this expression we get dot products of two cross products. They can be evaluated as

$$
\begin{aligned}
\left[\mathbf{k} \times \boldsymbol{\varepsilon}_\lambda(\mathbf{k})\right] \cdot \left[\mathbf{k}' \times \boldsymbol{\varepsilon}^*_{\lambda'}(\mathbf{k}')\right] &= \boldsymbol{\varepsilon}^*_{\lambda'}(\mathbf{k}') \cdot \left[\{\mathbf{k} \times \boldsymbol{\varepsilon}_\lambda(\mathbf{k})\} \times \mathbf{k}'\right] \\
&= -\boldsymbol{\varepsilon}^*_{\lambda'}(\mathbf{k}') \cdot \left[\mathbf{k}' \times \{\mathbf{k} \times \boldsymbol{\varepsilon}_\lambda(\mathbf{k})\}\right] \\
&= -\boldsymbol{\varepsilon}^*_{\lambda'}(\mathbf{k}') \cdot \left[\{\mathbf{k}' \cdot \boldsymbol{\varepsilon}_\lambda(\mathbf{k})\}\mathbf{k} - (\mathbf{k} \cdot \mathbf{k}')\boldsymbol{\varepsilon}_\lambda(\mathbf{k})\right] \\
&= (\mathbf{k} \cdot \mathbf{k}')\left[\boldsymbol{\varepsilon}_\lambda(\mathbf{k}) \cdot \boldsymbol{\varepsilon}^*_{\lambda'}(\mathbf{k}')\right] \qquad (1.52)
\end{aligned}
$$

The first step in the above equation is a (counter-clockwise) cyclic permutation as in the triple product $\mathbf{a} \cdot (\mathbf{b} \times \mathbf{c}) = \mathbf{c} \cdot (\mathbf{a} \times \mathbf{b})$. The third step expands the triple product as in $\mathbf{a} \times (\mathbf{b} \times \mathbf{c}) = (\mathbf{a} \cdot \mathbf{c})\mathbf{b} - (\mathbf{a} \cdot \mathbf{b})\mathbf{c}$. In the fourth step we use that $\mathbf{k}' = \pm\mathbf{k}$ and thus $\mathbf{k}' \cdot \boldsymbol{\varepsilon}_\lambda(\mathbf{k}) = \pm\mathbf{k} \cdot \boldsymbol{\varepsilon}_\lambda(\mathbf{k}) = 0$ according to Equation (1.19). The other three dot products of two cross products can be evaluated in a manner similar to Equation (1.52).

When working out Equation (1.51) we notice some similarity between this calculation and the one done earlier for $|\mathbf{E}|^2$, especially in the way the products of the exponentials are structured. We get four terms, each summed over $\mathbf{k}, \lambda, \mathbf{k}', \lambda'$. There are two types of exponentials among the four terms: those with the difference and those with the sum of k and k' in the exponent. We reorder Equation (1.51) with the terms involving the difference $k - k'$ first and insert the four dot products of two cross products as evaluated in Equation (1.52) in their appropriate terms

$$
\begin{aligned}
|\mathbf{B}|^2 = \frac{4\pi}{2V} \sum_{\mathbf{k},\mathbf{k}'\lambda\lambda'} \sqrt{\frac{1}{\omega_k \omega'_k}} \left[a_\lambda(\mathbf{k}) a^\dagger_{\lambda'}(\mathbf{k}') e^{i(k-k')x} (\mathbf{k} \cdot \mathbf{k}') \{\boldsymbol{\varepsilon}_\lambda(\mathbf{k}) \cdot \boldsymbol{\varepsilon}^*_{\lambda'}(\mathbf{k}')\} + \overline{\mathrm{h.c.}}\,\right] \\
- \frac{4\pi}{2V} \sum_{\mathbf{k},\mathbf{k}',\lambda,\lambda'} \sqrt{\frac{1}{\omega_k \omega'_k}} \left[a_\lambda(\mathbf{k}) a_{\lambda'}(\mathbf{k}') e^{i(k+k')x} (\mathbf{k} \cdot \mathbf{k}') \{\boldsymbol{\varepsilon}_\lambda(\mathbf{k}) \cdot \boldsymbol{\varepsilon}_{\lambda'}(\mathbf{k}')\} + \overline{\mathrm{h.c.}}\,\right]
\end{aligned}
\qquad (1.53)
$$

We exchange the integration over all of space with the summations, and use Equation (1.48) as before to find that the double sum over k and k′ reduces to a single sum over $\mathbf{k}$. The time-dependent exponentials cancel in the two terms with $\mathbf{k}' = \mathbf{k}$ but not in the two terms with $\mathbf{k}' = -\mathbf{k}$. Note that the factor $\mathbf{k} \cdot \mathbf{k}'$ that appears in Equation (1.53) is equal to $\pm\mathbf{k}^2 = \pm\omega_k^2$ depending upon the relative signs of $\mathbf{k}$ and $\mathbf{k}'$. We get

$$
\begin{aligned}
\int d^3\mathbf{x}\, \mathbf{B}^2 = 2\pi \sum_{\mathbf{k},\lambda,\lambda'} \frac{1}{\omega_k} \left[a_\lambda(\mathbf{k}) a^\dagger_{\lambda'}(\mathbf{k}) \{\boldsymbol{\varepsilon}_\lambda(\mathbf{k}) \cdot \boldsymbol{\varepsilon}^*_{\lambda'}(\mathbf{k})\} + \overline{\mathrm{h.c.}}\,\right] (+\omega_k^2) \\
-2\pi \sum_{\mathbf{k},\lambda,\lambda'} \frac{1}{\omega_k} \left[a_\lambda(\mathbf{k}) a_{\lambda'}(-\mathbf{k})\, e^{-2i\omega_k t} \{\boldsymbol{\varepsilon}_\lambda(\mathbf{k}) \cdot \boldsymbol{\varepsilon}_{\lambda'}(-\mathbf{k})\} + \overline{\mathrm{h.c.}}\,\right] (-\omega_k^2)
\end{aligned}
\qquad (1.54)
$$

Note the factor $+\omega_k^2$ in the first sum and $-\omega_k^2$ in the second sum.

When we sum Equation (1.49) and Equation (1.54) we see that the second sum in Equation (1.49) cancels the second sum in Equation (1.54) as mentioned earlier. This is good because the cancellation removes the time dependence present in $|\mathbf{E}|^2$ and $|\mathbf{B}|^2$ so that the energy E_{em} is time-independent as it should be for an electromagnetic field in vacuum. The cancellation also gets rid of the troublesome terms involving the dot product of polarization vectors whose factors have too few or too many complex conjugations and are evaluated at different values of their momentum. The first sums in Equation (1.49) and Equation (1.54) are equal so we get a factor 2. Inserting the factor $1/(8\pi)$ from Equation (1.20) we get, setting $E_{\text{em}} = H$

$$H = \tfrac{1}{2} \textstyle\sum_{\mathbf{k},\lambda} \omega_k \left[a_\lambda(\mathbf{k})a_\lambda^\dagger(\mathbf{k}) + a_\lambda^\dagger(\mathbf{k})a_\lambda(\mathbf{k})\right] \tag{1.55}$$

$$= \tfrac{1}{2} \textstyle\sum_{\mathbf{k},\lambda} \omega_k \left[2a_\lambda^\dagger(\mathbf{k})a_\lambda(\mathbf{k}) + 1\right] \tag{1.56}$$

$$= \sum_{\mathbf{k},\lambda} \left[N_\lambda(\mathbf{k}) + \tfrac{1}{2}\right] \omega_k \tag{1.57}$$

where we have used the commutation relation in Equation (1.31) in Equation (1.56) and introduced the number operator $N_\lambda(\mathbf{k})$ from Equation (1.35) in Equation (1.57).

This is a remarkable result for several reasons. First, it shows that the energy of the electromagnetic field is indeed quantized so that we may speak of 'photons'. Second, it shows that the quantum of energy is ω_k ($\hbar\omega_k$ really) where $\omega_k = |\mathbf{k}|$. The quantities where $\mathbf{k}$ and ω were first introduced in the plane wave solutions of the wave Equation (1.21) and they derive their physical meaning from that expression: ω is the angular frequency. Thus we find the relation $E = \hbar\omega$, the equation Planck wrote down for black-body radiation. And third, it has the remarkable consequence that even if there are no photons ($N_\lambda(\mathbf{k}) = 0$ for all $\mathbf{k}$ and λ) the energy is not zero but in fact infinite! This is annoying to say the least and a sign of serious trouble ahead. In the meantime we give it a name: 'zero point energy' and measure all energies relative to it. This is the second time that one encounters this kind of nonsense; the first time was in the treatment of the Harmonic Oscillator to which the current trouble can be traced. So it's really only one problem, some consolation!

The analogy between the above result for the energy spectrum of electromagnetic energy and the energy spectrum of the Harmonic Oscillator is the justification for speaking in terms of 'oscillators' present in an electromagnetic field when giving an elementary derivation of Black Body Radiation in introductory physics. But it's no more than a manner of speaking.

We now see why the choice of pre-factor in Equation (1.29), Equation (1.30) and Equation (1.41) is a good one even though it appeared to be rather arbitrary. With the choice of pre-factors in those relations we obtain the expression for the Hamiltonian Equation (1.57). Had we made a different choice then additional factors would have appeared in Equation (1.57) in disagreement with experiment, for example, black-body radiation.

1.2.3 Momentum

The momentum of the electromagnetic field in a vacuum is given by the classical expression

$$\mathbf{P}_{\mathrm{em}} = \frac{1}{4\pi}\int d^3\mathbf{x}\,\mathbf{E}\times\mathbf{B}^* \tag{1.58}$$

The complex conjugate of **B** appears so that **P** is parallel to **k**. We replace the complex conjugate by the Hermitian conjugate but in the present case **B** is Hermitian so the † does not matter. We calculate the $\mathbf{E}\times\mathbf{B}^\dagger$ using Equation (1.44) and Equation (1.45)

$$\begin{aligned}\mathbf{E}\times\mathbf{B}^\dagger &= -\frac{4\pi}{V}\sum_{\mathbf{k},\lambda}\sqrt{\frac{\omega_k}{2}}\left[a_\lambda(\mathbf{k})e^{ikx}\boldsymbol{\varepsilon}_\lambda(\mathbf{k}) - a_\lambda^\dagger(\mathbf{k})e^{-ikx}\boldsymbol{\varepsilon}_\lambda^*(\mathbf{k})\right]\\ &\quad\times\sum_{\mathbf{k}',\lambda'}\sqrt{\frac{1}{2\omega_k'}}\left[a_{\lambda'}^\dagger(\mathbf{k}')e^{-ik'x}\mathbf{k}'\times\boldsymbol{\varepsilon}_{\lambda'}^*(\mathbf{k}') - a_{\lambda'}(\mathbf{k}')e^{ik'x}\mathbf{k}'\times\boldsymbol{\varepsilon}_{\lambda'}(\mathbf{k}')\right]\end{aligned} \tag{1.59}$$

When working out this expression we get a cross product of a vector and a cross product of two other vectors. They can be evaluated as

$$\begin{aligned}\boldsymbol{\varepsilon}_\lambda(\mathbf{k})\times\left[\mathbf{k}'\times\boldsymbol{\varepsilon}_{\lambda'}^*(\mathbf{k}')\right] &= \left[\boldsymbol{\varepsilon}_\lambda(\mathbf{k})\cdot\boldsymbol{\varepsilon}_{\lambda'}^*(\mathbf{k}')\right]\mathbf{k}' - \left[\varepsilon_\lambda(\mathbf{k})\cdot\mathbf{k}'\right]\varepsilon_{\lambda'}^*\\ &= \left[\boldsymbol{\varepsilon}_\lambda(\mathbf{k})\cdot\boldsymbol{\varepsilon}_{\lambda'}^*(\mathbf{k}')\right]\mathbf{k}'\end{aligned} \tag{1.60}$$

The first step in the above equation expands the triple product as in $\mathbf{a}\cdot(\mathbf{b}\times\mathbf{c}) = \mathbf{c}\cdot(\mathbf{a}\times\mathbf{b})$. In the second step we used that $\mathbf{k}' = \pm\mathbf{k}$ and thus $\boldsymbol{\varepsilon}_\lambda(\mathbf{k})\cdot\mathbf{k}' = \boldsymbol{\varepsilon}_\lambda\cdot(\pm\mathbf{k}) = 0$ according to Equation (1.19). The other three cross products of a vector and a cross product can be evaluated in a manner similar to Equation (1.60). When working out Equation (1.60) we notice some similarity between this calculation and the one done earlier for $|\mathbf{E}|^2$ and $|\mathbf{B}|^2$, especially in the way the products of the exponentials are structured. We get four terms, each summed over $\mathbf{k}, \lambda, \mathbf{k}', \lambda'$. There are two types of exponentials among the four terms: those with the difference and those with the sum of k and k' in the exponent. We reorder Equation (1.59) with the terms involving the difference $k - k'$ first and insert the four products of the type Equation (1.60) in their appropriate terms to get

$$\mathbf{E}\times\mathbf{B}^{\dagger} = -\frac{4\pi}{2V}\sum_{\mathbf{k},\mathbf{k}'\lambda\lambda'}\left[a_{\lambda}(\mathbf{k})a^{\dagger}_{\lambda'}(\mathbf{k}')e^{i(k-k')x}\{\boldsymbol{\varepsilon}_{\lambda}(\mathbf{k})\cdot\boldsymbol{\varepsilon}^{*}_{\lambda'}(\mathbf{k}')\}\,\mathbf{k}' + \overline{\text{h.c.}}\right]$$
$$-\frac{4\pi}{2V}\sum_{\mathbf{k},\mathbf{k}',\lambda,\lambda'}\left[a_{\lambda}(\mathbf{k})a_{\lambda'}(\mathbf{k}')e^{i(k+k')x}\{\boldsymbol{\varepsilon}_{\lambda}(\mathbf{k})\cdot\boldsymbol{\varepsilon}_{\lambda'}(\mathbf{k}')\}\,\mathbf{k}' + \overline{\text{h.c.}}\right] \tag{1.61}$$

We exchange the integration over all of space with the summations and use Equation (1.48) as before to find that the double sum over $\mathbf{k}$ and $\mathbf{k}'$ reduces to a single sum over $\mathbf{k}$. The time-dependent exponentials cancel in the two terms with $\mathbf{k}' = \mathbf{k}$ but not in the two terms with $\mathbf{k}' = -\mathbf{k}$

$$\int d^3\mathbf{x}\,\mathbf{E}\times\mathbf{B}^{\dagger} = 2\pi\sum_{\mathbf{k},\lambda,\lambda'}\left[a_{\lambda}(\mathbf{k})a^{\dagger}_{\lambda'}(\mathbf{k})\{\boldsymbol{\varepsilon}_{\lambda}(\mathbf{k})\cdot\boldsymbol{\varepsilon}^{*}_{\lambda'}(\mathbf{k})\}\,\mathbf{k} + \overline{\text{h.c.}}\right]\mathbf{k}$$
$$+2\pi\sum_{\mathbf{k},\lambda,\lambda'}\left[a_{\lambda}(\mathbf{k})a_{\lambda'}(-\mathbf{k})\,e^{-2i\omega_k t}\{\boldsymbol{\varepsilon}_{\lambda}(\mathbf{k})\cdot\boldsymbol{\varepsilon}_{\lambda'}(-\mathbf{k})\}\,\mathbf{k} + \overline{\text{h.c.}}\right] \tag{1.62}$$

Note that all terms are proportional to $\mathbf{k}$ as they should be because $\mathbf{E}\times\mathbf{B}^{\dagger}$ should be parallel to $\mathbf{k}$ for each value of $\mathbf{k}$. The second sum is zero because its terms are antisymmetric under the exchange $\mathbf{k}\to-\mathbf{k}$, while positive and negative values of $\mathbf{k}$ appear equally. This is good because the time dependence of $\mathbf{P}_{\text{em}}$ is removed as expected because the momentum of an electromagnetic field in a vacuum is conserved. We also get rid of the troublesome terms involving the dot product of polarization vectors whose factors have too few or too many complex conjugations and are evaluated at different values of their momentum. To show that the second sum in Equation (1.62) is zero, we split the sum over $\mathbf{k}$ into two equal pieces, each with a factor $\frac{1}{2}$ in front, and then replace $\mathbf{k}\to-\mathbf{k}'$ in the first sum

$$\sum_{\mathbf{k},\lambda,\lambda'} a_{\lambda}(\mathbf{k})a_{\lambda'}(-\mathbf{k})\,e^{-2i\omega_k t}\{\boldsymbol{\varepsilon}_{\lambda}(\mathbf{k})\cdot\boldsymbol{\varepsilon}_{\lambda'}(-\mathbf{k})\}\,\mathbf{k}$$
$$= \tfrac{1}{2}\textstyle\sum_{\mathbf{k},\lambda,\lambda'} a_{\lambda}(\mathbf{k})a_{\lambda'}(-\mathbf{k})\,e^{-2i\omega_k t}\{\boldsymbol{\varepsilon}_{\lambda}(\mathbf{k})\cdot\boldsymbol{\varepsilon}_{\lambda'}(-\mathbf{k})\}\,\mathbf{k}\,+$$
$$\tfrac{1}{2}\textstyle\sum_{\mathbf{k},\lambda,\lambda'} a_{\lambda}(\mathbf{k})a_{\lambda'}(-\mathbf{k})\,e^{-2i\omega_k t}\{\boldsymbol{\varepsilon}_{\lambda}(\mathbf{k})\cdot\boldsymbol{\varepsilon}_{\lambda'}(-\mathbf{k})\}\,\mathbf{k}$$
$$= \tfrac{1}{2}\textstyle\sum_{\mathbf{k}',\lambda,\lambda'} a_{\lambda}(-\mathbf{k}')a_{\lambda'}(\mathbf{k}')\,e^{-2i\omega_k t}\{\boldsymbol{\varepsilon}_{\lambda}(-\mathbf{k}')\cdot\boldsymbol{\varepsilon}_{\lambda'}(\mathbf{k}')\}\,(-\mathbf{k}')\,+$$
$$\tfrac{1}{2}\textstyle\sum_{\mathbf{k},\lambda,\lambda'} a_{\lambda}(\mathbf{k})a_{\lambda'}(-\mathbf{k})\,e^{-2i\omega_k t}\{\boldsymbol{\varepsilon}_{\lambda}(\mathbf{k})\cdot\boldsymbol{\varepsilon}_{\lambda'}(-\mathbf{k})\}\,\mathbf{k} \tag{1.63}$$

Inspection shows that the last two sums cancel each other because of the minus sign in front of $\mathbf{k}'$ in the first of the two sums. The factors multiplying $-\mathbf{k}'$ and $\mathbf{k}$ in the last two sums are equal because $a_{\lambda}(\mathbf{k})$ and $a_{\lambda'}(-\mathbf{k})$ commute for all $\mathbf{k}, \mathbf{k}', \lambda, \lambda'$, see Equation (1.32), and in executing the double sum over λ and λ' it does not matter in which order one does this. The h.c. term in the second sum in Equation (1.62) is also zero, of course, because it is the Hermitian conjugate of the one we just considered apart from the order of the operators, but they commute as well. The symmetry argument (luckily) does not hold for the first sum in Equation (1.62) because that sum involves the product of $a_\lambda(\mathbf{k})a^\dagger_{\lambda'}(\mathbf{k})$ and its h.c. These do not commute, see Equation (1.31), so the trick employed in Equation (1.63) does not work for the first sum.

Using Equation (1.50) for the dot product of the polarization vectors and inserting the factor $1/(4\pi)$ from Equation (1.58) we get

$$\mathbf{P}_{\text{em}} = \tfrac{1}{2}\textstyle\sum_{\mathbf{k},\lambda}\left[a_\lambda(\mathbf{k})a^\dagger_\lambda(\mathbf{k}) + a^\dagger_\lambda(\mathbf{k})a_\lambda(\mathbf{k})\right]\mathbf{k} \tag{1.64}$$

$$= \tfrac{1}{2}\textstyle\sum_{\mathbf{k},\lambda}\, a^\dagger_\lambda(\mathbf{k})a_\lambda(\mathbf{k}) + 1\big]\,\mathbf{k} \tag{1.65}$$

$$= \sum_{\mathbf{k},\lambda}\left[N_\lambda(\mathbf{k}) + \tfrac{1}{2}\right]\mathbf{k} \tag{1.66}$$

$$= \sum_{\mathbf{k},\lambda} N_\lambda(\mathbf{k})\,\mathbf{k} \tag{1.67}$$

where we have used the commutation relations Equation (1.31) in Equation (1.64) and introduced the number operator $N_\lambda(\mathbf{k})$ from Equation (1.35) in Equation (1.65). The 'zero point momentum' evident in Equation (1.66) sums to zero if $\mathbf{k}$ has no preferred direction for an electromagnetic field in a vacuum, a reasonable assumption. We evade the zero point momentum, contrary to the case for the energy, see Equation (1.57).

The result in Equation (1.67) is equally remarkable as the result in Equation (1.57) for the same reasons. It shows that the momentum of the electromagnetic field is quantized, confirming that we may speak of 'photons'. The quantum of momentum is $\mathbf{k}$ ($\hbar\mathbf{k}$ really) where $\mathbf{k}$ was first introduced in the plane wave solutions of the wave Equation (1.21). The vector $\mathbf{k}$ derives its physical meaning from that expression: $|\mathbf{k}|$ is the wave vector with magnitude $2\pi/\lambda$. We conclude that the momentum $\hbar\mathbf{k}$ of a photon is equal to $\hbar 2\pi/\lambda$ or $p = h/\lambda$, the deBroglie relation. We can use the Einstein relation $m^2 = E^2 - \mathbf{p}^2$ to calculate the mass of the photon. Using $E = \hbar\omega_k$ and $\mathbf{p} = \hbar\mathbf{k}$ we find $m^2 = 0$. This is as expected because a 'particle' can travel with the speed of light (in a vacuum) only if it is massless.

Also here we see that the choice of pre-factor in Equation (1.29) and Equation (1.30) is a good one because it leads again to a result in agreement with experiment.

1.2.4 Polarization and Spin

The polarization factors $\boldsymbol{\varepsilon}_\lambda(\mathbf{k})$ and its complex conjugate define the polarization as discussed earlier. A photon with momentum $\mathbf{k}$ along the z-axis that is linearly polarized along the x-axis is represented by $a_1^\dagger(\mathbf{k})\,|0\rangle$, in analogy with a plane wave in classical electromagnetism. Likewise a photon linearly polarized along the y-axis is represented by $a_2^\dagger(\mathbf{k})\,|0\rangle$. Here the unit vectors $\boldsymbol{\varepsilon}_1$ and $\boldsymbol{\varepsilon}_2$ are chosen along the x-axis and the y-axis respectively and they form with $\mathbf{k}/\omega_k$ a right-handed triplet of (real) unit vectors in the order $(\boldsymbol{\varepsilon}_1, \boldsymbol{\varepsilon}_2, \mathbf{k}/\omega_k)$. When discussing linearly polarized photons, we can and do omit the complex conjugate symbol over $\boldsymbol{\varepsilon}_\lambda(\mathbf{k})$. Such photons are called transverse photons.

To represent circularly polarized photons we introduce

$$\begin{aligned} a_R(\mathbf{k}) &= a_1(\mathbf{k}) + ia_2(\mathbf{k}) \\ a_L(\mathbf{k}) &= a_1(\mathbf{k}) - ia_2(\mathbf{k}) \end{aligned} \tag{1.68}$$

We will show that a right-handed (RH) photon is created by $|\mathbf{k}, R\rangle = a_R^\dagger(\mathbf{k})|0\rangle$ and that a left-handed (LH) photon is created by $|\mathbf{k}, L\rangle = a_L^\dagger(\mathbf{k})|0\rangle$. Creation operators in the expression for $\mathbf{A}(\mathbf{x}, t)$ are accompanied by a factor $\exp(-ikx)$. If we fix the position $\mathbf{x}$ and let t vary we get a factor $\exp(i\omega t)$. We separate the real and imaginary parts of the resulting expression

$$\begin{aligned} a_R^\dagger(\mathbf{k})e^{i\omega t}|0\rangle &= (a_1 + ia_2)^\dagger e^{i\omega t}|0\rangle \\ &= (a_1^\dagger - ia_2^\dagger)(\cos\omega t + i\sin\omega t)|0\rangle \\ &= (a_1^\dagger\cos\omega t + a_2^\dagger\sin\omega t)|0\rangle + i(a_1^\dagger\sin\omega t - a_2^\dagger\cos\omega t)|0\rangle \end{aligned} \tag{1.69}$$

Consider first the real part of Equation (1.69). We plot in Figure 1.1 on an x, y coordinate system the *coefficient* of $a_1^\dagger$ plotted along the x-axis and the *coefficient* of $a_2^\dagger$ plotted along the y-axis (the respective directions of polarization related to $a_1^\dagger$ and $a_2^\dagger$). It is seen that the point defined by these two coefficients lies on a circle and that the point moves in a counter-clockwise direction (in the direction from the x-axis to the y-axis) around the circle as time increases. The latter statement can be checked by substituting $\omega t = 0$ (A in Figure 1.1) and $\omega t = \pi/2$ (B in Figure 1.1) in Equation (1.69) and following the position of the point along the circle from A to B. The same counter-clockwise direction of movement of the point is specified by the imaginary part of Equation (1.69). This is consistent with the subscript R of the coefficient a_R. The photon moves in the direction of $\mathbf{k}$, in the figure out of the paper toward the reader, and the counter-clockwise movement of the point corresponds to a RH movement relative to $\mathbf{k}$ in the manner defined by the right-hand rule of the cross product of two vectors. In a similar manner we can show that a LH photon is represented by $|\mathbf{k}, L\rangle = a_L^\dagger(\mathbf{k})|0\rangle$. For our discussion we have selected the term in $\mathbf{A}$ that creates a photon. One can (and should) verify that the arguments also hold for the terms that annihilate a photon.

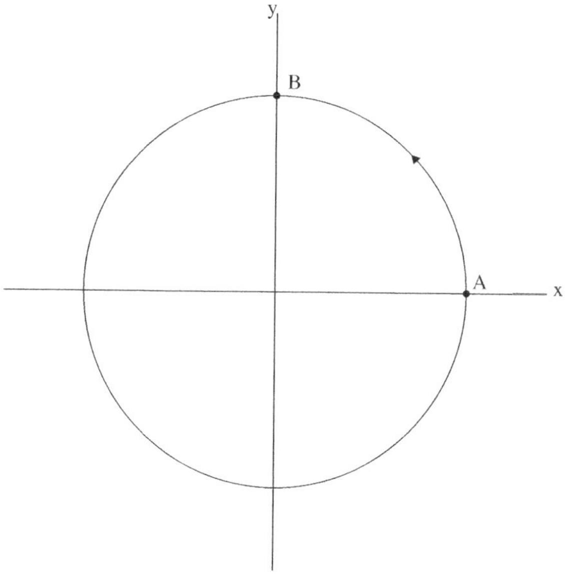

Figure 1.1 Illustration of left-handed and right-handed polarization

In optics the naming convention is the opposite: what we call a RH photon is called LH in optics because in optics one looks into the oncoming light when determining the sense of its circular polarization.

To discuss the polarization of the electric field **E** we introduce in analogy with Equation (1.68)

$$\boldsymbol{\varepsilon}_R(\mathbf{k}) = \boldsymbol{\varepsilon}_1(\mathbf{k}) + i\boldsymbol{\varepsilon}_2(\mathbf{k})$$
$$\boldsymbol{\varepsilon}_L(\mathbf{k}) = \boldsymbol{\varepsilon}_1(\mathbf{k}) - i\boldsymbol{\varepsilon}_2(\mathbf{k}) \tag{1.70}$$

Note that the $\boldsymbol{\varepsilon}_R$ and $\boldsymbol{\varepsilon}_L$ are complex; that is the reason we distinguished between polarization vectors and their complex conjugates throughout. A RH polarized electric field can be represented by a RH polarized $\mathbf{A}(\mathbf{x}, t)$ because according to Equation (1.17) **E** and **A** differ by a multiplicative (complex constant) and such a constant does not change the relative handedness of **E** and **A**. When we fix the position **x** and let t vary, we get for a RH polarized **A**

$$\mathbf{A}(\mathbf{x}, t) \propto \varepsilon_R^* e^{i\omega t} = (\boldsymbol{\varepsilon}_1^* - i\boldsymbol{\varepsilon}_2^*)(\cos\omega t + i\sin\omega t)$$
$$= (\boldsymbol{\varepsilon}_1\cos\omega t + \boldsymbol{\varepsilon}_2\sin\omega t) + i(\boldsymbol{\varepsilon}_1\sin\omega t - \boldsymbol{\varepsilon}_2\cos\omega t) \tag{1.71}$$

This equation has the same structure as Equation (1.69) so it represents a counter-clockwise rotating vector potential **A** and leads to a counter-clockwise rotating **E**. They rotate in a RH sense with respect to the momentum **k**. The subscript R on ε_R also here correctly describes the polarization of **A** and **E**. In a similar manner one can show that a LH polarized **A** is given by

$$\mathbf{A}(\mathbf{x}, t) \propto \varepsilon_L^* e^{i\omega t} \tag{1.72}$$

For our discussion we have selected the Hermitian conjugate terms of the expression for **A**. One can (and should) verify that the arguments hold for the other term too. We will return to circular polarization in Section 1.6.

The photon has spin 1. This follows from the fact that A_μ is a four-vector whose fourth component (the scalar potential ϕ) is identically zero in the Coulomb gauge. The remaining three components are constrained by the Lorenz condition in Equation (1.19) leaving only two independent components. Without getting too formal in proving this, we assert that these two components correspond to the two z-components $s_z = +1$ and $s_z = -1$ of a spin 1 particle and not a $s = \frac{1}{2}$ particle as one might initially think. These two possibilities correspond to RH and LH circular polarizations respectively.

Another characterization of the handedness of the circular polarization involves helicity. One defines the helicity h of a particle with momentum **p** and spin s as

$$h = \frac{\mathbf{s} \cdot \mathbf{p}}{|\mathbf{s}||\mathbf{p}|} \tag{1.73}$$

It is seen that a RH photon with $s_z = +1$ has $h = +1$, while a LH photon with $s_z = -1$ has $h = -1$ if the axis of quantization is chosen along **p**.

The absence of the $s_z = 0$ component is ultimately a consequence of the fact that the photon is massless: the Lorenz and thus the Coulomb Gauge can only be implemented for massless photons. Thus Equation (1.19) is true only for massless photons (classically: waves that in a vacuum travel at the speed of light) and only then is $\boldsymbol{\varepsilon}_\lambda$ perpendicular to **k**. This can be seen by tracing the derivations of subsection 1.1. If the electromagnetic interaction is described in a relativistically correct manner (we will not do this here) one finds that the photon acquires a mass in some cases. We say that the photon is 'off the mass shell' because it does not satisfy the Einstein relation $E^2 - \mathbf{p}^2 = m^2$ (with $m = 0$ in this case) as an 'on the mass shell' particle will. When this happens it acquires a $s_z = 0$ component! Such photons are called longitudinal or scalar photons.

There is another example of a massless particle, the mediator of the gravitational interaction, the graviton whose spin is 2 with only two values for its z-component $s_z = \pm 2$.

Note that when in classical electromagnetism a plane wave is called transversely polarized, it is called longitudinally polarized in quantum physics. This is confusing! In Table 1.1 we compare the classical and quantum physical properties of electromagnetic radiation.

Table 1.1 Comparison of terminology for photons

Classical	Quantum Physical
Wave	Photon
Velocity c	Mass zero
RH and LH transverse polarization	Helicity is $+1$ and -1
No longitudinal polarization	No $s_z = 0$

The number operator $N_\lambda(\mathbf{k})$ does not commute with **A, E** or **B**. This is because the number operator does not commute with a_λ and $a_\lambda^\dagger$ as can be seen from equations (1.36) and (1.37) and because Equation (1.30), Equation (1.44) and Equation (1.45) for **A, E** and **B** have terms proportional to either a_λ or $a_\lambda^\dagger$. The non-zero commutators lead to uncertainty relations between $N_\lambda(\mathbf{k})$ and **E** and **B**. When the operators corresponding to two observables do not commute, an uncertainty relation results for their eigenvalues. The prototypes for this are momentum and position or energy and time, see Equation (1.7), leading to the Heisenberg uncertainty relations. So if the eigenvalue of $N_\lambda(\mathbf{k})$ is specified (so we know exactly how many photons are present in the system) the values of **E** and **B** (and **A**) are uncertain. This leads to the strange conclusion that the vacuum (which has exactly zero photons) has uncertain values of its electric and magnetic fields. One can show that while $\langle 0|\mathbf{E}|0\rangle = 0$ and $\langle 0|\mathbf{B}|0\rangle = 0$ the values of $\langle 0|\mathbf{E}^2|0\rangle$ and $\langle 0|\mathbf{B}^2|0\rangle$ are infinite and therefore the standard deviations squared σ_E^2 and σ_B^2 are infinite and the electromagnetic energy $E_{\text{em}} = \infty$ according to Equation (1.20). Compare this with the zero point energy problem of the harmonic oscillator and the electromagnetic field. This means trouble! In the classical domain with well defined electric and magnetic fields the number of photons is *extremely* large, and averages over finite volumes of space lead to meaningful quantities. In the quantum physical domain we deal with an exactly known but *very small* number of photons $(0, 1, 2, \ldots)$ and the electric and magnetic field have uncertain values (but we do not care to know them).

1.2.5 Hamiltonian

The Hamiltonian for a charged particle in an electromagnetic field is given by the sum of two parts, one given by Equation (1.1), the other by Equation (1.20) as

$$H = \frac{1}{2m}(\mathbf{p} - e\mathbf{A})^2 + e\phi + \frac{1}{8\pi}\int d^3\,\mathbf{x}(\mathbf{E}^2 + \mathbf{B}^2) \tag{1.74}$$

If we expand the first part we get

$$H = \frac{\mathbf{p}^2}{2m} + e\phi + \frac{1}{8\pi}\int d^3\,\mathbf{x}(\mathbf{E}^2 + \mathbf{B}^2) - \frac{e}{2m}\mathbf{p}\cdot\mathbf{A} - \frac{e}{2m}\mathbf{A}\cdot\mathbf{p} + \frac{e^2}{2m}\mathbf{A}^2 \tag{1.75}$$

where we have written 'large' terms first. The value of e is given by $e^2 = \alpha \approx 1/137$ so the last three terms can be considered 'small' in comparison with the first three terms. We note that terms with $\mathbf{p} \cdot \mathbf{A}$ and $\mathbf{A} \cdot \mathbf{p}$ can be combined. Using a test function $\psi(\mathbf{x})$ we have

$$\mathbf{p} \cdot \mathbf{A}\,\psi = \frac{1}{i}\nabla \cdot \mathbf{A}\,\psi = \frac{1}{i}(\nabla \cdot \mathbf{A})\,\psi + \frac{1}{i}\mathbf{A} \cdot \nabla\psi = \mathbf{A} \cdot \mathbf{p}\,\psi \tag{1.76}$$

We used Equation (1.5) in the first equality, the product rule of calculus in the second, and Equation (1.18) and Equation (1.5) for the third. Again we see the advantage of using the Coulomb Gauge. Because the relation in Equation (1.76) is true for any ψ we conclude that $[\mathbf{p}, \mathbf{A}] = 0$ in the Coulomb Gauge and the Hamiltonian Equation (1.75) becomes with operators shown explicitly

$$H = -\frac{\nabla^2}{2m} + e\phi + \sum_{\mathbf{k},\lambda}\left[N_\lambda(\mathbf{k}) + \frac{1}{2}\right]\omega_k - \frac{e}{im}\mathbf{A} \cdot \nabla + \frac{e^2}{2m}\mathbf{A}^2 \tag{1.77}$$

The physical consequences of this Hamiltonian can be inferred by using time-dependent perturbation theory with the last two terms the time-dependent perturbation. The first three terms correspond to an unperturbed Hamiltonian H_0 with

$$H_0 = -\frac{\nabla^2}{2m} + e\phi + \sum_{\mathbf{k},\lambda}\left[N_\lambda(\mathbf{k}) + \frac{1}{2}\right]\omega_k \tag{1.78}$$

We will review time-dependent perturbation theory in the next subsection.

1.3 TIME-DEPENDENT PERTURBATION THEORY

This will be a brief review because time-dependent perturbation theory will have been discussed in detail in your favorite quantum physics book. Our treatment is not relativistically correct as is shown explicitly by the appearance of the time variable without the equivalent appearance of the three space variables. The time dependence leads to energy conservation explicit in the amplitude for a process without the equivalent three-momentum conservation. Special relativity requires that all four components of the energy-momentum four-vector appear without preference for any one of them. We will see that energy-momentum conservation can be violated in a process with an appeal to Heisenberg's uncertainty relations. In a relativistically correct formulation, energy-momentum conservation is respected in all parts of a process, but the photon can sometimes have non-zero mass and be off the mass shell.

We assume that we can write the Hamiltonian in two pieces, a 'large' time-independent one H_0 and a 'small' time-dependent one $\lambda H_1(t) = \lambda V \exp(\pm i\omega t)$ with all time dependence in the exponential ('Harmonic Perturbation') so V is a time-independent operator. The parameter λ is introduced to keep track of the order of magnitude in the perturbation expansion of the solution to the Schrödinger equation. Thus we have

$$H = H_0 + \lambda H_1(t) = H_0 + \lambda V e^{\pm i\omega t} \tag{1.79}$$

We assume that the eigenkets $|\phi_n\rangle$ and eigenvalues $E_n^{(0)}$ of H_0 are known, that is we assume that the problem is solved for $\lambda = 0$

$$H_0|\phi_n\rangle = E_n^{(0)}|\phi_n\rangle \tag{1.80}$$

and that the kets $|\phi_n\rangle$ are orthonormal

$$\langle\phi_m|\phi_n\rangle = \delta_{mn} \tag{1.81}$$

and form a complete set. Let's simplify the notation by letting $E_n^{(0)} \to E_n$ and $|\phi_n\rangle \to |n\rangle$. The most general eigenket of H_0 (and of the full Hamiltonian H with $\lambda = 0$) is then a superposition of eigenkets

$$|\psi\rangle = \sum_n c_n^{(0)}|n\rangle e^{-iE_nt} \tag{1.82}$$

where the $c_n^{(0)}$ are constant. We find the most general eigenkets and eigenvalues of the full Hamiltonian Equation (1.79) with $\lambda \neq 0$ using a linear superposition as in Equation (1.82) but with the coefficients functions of time

$$|\psi\rangle = \sum_n c_n(t)|n\rangle e^{-iE_nt} \tag{1.83}$$

This should work because the eigenkets $|n\rangle$ form a complete set. When this expression is substituted in the Schrödinger equation we get an equation for the coefficients $c_n(t)$

$$\sum_n c_n(t)\lambda H_1|n\rangle e^{-iE_nt} = -\frac{1}{i}\sum_n \dot{c}_n(t)|n\rangle e^{-iE_nt} \tag{1.84}$$

In order to keep track of terms of the same order in λ, we set

$$c_n(t) = c_n^{(0)} + \lambda c_n^{(1)}(t) + \lambda^2 c_n^{(2)}(t) + \cdots \tag{1.85}$$

Note that for $\lambda = 0$ the new $c_n(t)$ are equal to the old time-independent $c_n^{(0)}$ of Equation (1.82) as they should be. Substitution of Equation (1.85) in Equation (1.84) and equating terms with the same power of λ in the resulting equation (as the equation must be true if λ is varied) we get the iterative set of equations

$$0 = -\frac{1}{i}\sum_n \dot{c}_n^{(0)}\,|n\rangle\, e^{-iE_nt} \qquad \text{for } \lambda^0 \tag{1.86}$$

$$\sum_n c_n^{(0)} H_1 |n\rangle\, e^{-iE_nt} = -\frac{1}{i}\sum_n \dot{c}_n^{(1)}\,|n\rangle\, e^{-iE_nt} \qquad \text{for } \lambda^1 \tag{1.87}$$

$$\sum_n c_n^{(1)}(t) H_1 |n\rangle e^{-iE_nt} = -\frac{1}{i}\sum_n \dot{c}_n^{(2)}\,|n\rangle\, e^{-iE_nt} \qquad \text{for } \lambda^2 \tag{1.88}$$

$$\cdots \tag{1.89}$$

and so on for larger powers of λ. The structure of Equation (1.87), Equation (1.88) and Equation (1.89) can be seen. An equation of order λ^{n+1} can only be solved if the preceding equation of order λ^n is solved first. This requires an iterative method that we might call a 'Bootstrap'. Fortunately, we rarely go beyond the first non-zero result and, with rare exceptions, physicists never prove that the resulting series expansion converges. We solve for the coefficients $\dot{c}_n^{(f)}$ by multiplying Equation (1.86), Equation (1.87), Equation (1.88), $\cdots$ by $\langle f|$ from the left to find

$$\dot{c}_f^{(0)}(t) = 0 \quad \text{or} \quad c_f^{(0)} = \text{constant} \tag{1.90}$$

$$\lambda\dot{c}_f^{(1)}(t) = -i\sum_n c_n^{(0)} \langle f|\lambda H_1|n\rangle\, e^{i(E_f-E_n)t} \tag{1.91}$$

$$\lambda^2\dot{c}_f^{(2)}(t) = -i\sum_n \lambda c_n^{(1)} \langle f|\lambda H_1|n\rangle\, e^{i(E_f-E_n)t} \tag{1.92}$$

$$\cdots \tag{1.93}$$

Note that we multiplied the left and right side of each equation by an appropriate power of λ in accordance with the power of λ found in the terms in Equation (1.85). Thus far our results are general.

We now specialize to the case where the system is in the state $|i\rangle$ at some time t_1 in the past, and we are asking for the amplitude for the system to be in the state $|f\rangle$ at a later time t_2. We therefore set at t_1 all $c_n^{(m)}(t_1) = 0$ except that we set $c_i^{(0)} = 1$. The relation in Equation (1.85) shows that at t_1 all $c_n(t_1) = 0$ except that $c_i(t_1) = 1$, the 'initial condition'. The relation in Equation (1.82) shows that the system's state is $|\psi\rangle = |i\rangle$ as desired. To find the amplitude for the system to be in the state $|f\rangle$ at t_2 we project out the ket $|f\rangle$ from $|psi\rangle$ by calculating, using Equation (1.85), $c_f(t_2) = c_f^{(0)}(t_2) + \lambda c_f^{(1)}(t_2) + \lambda^2 c_f^{(2)}$

$(t_2) + \cdots$ where $c_f^{(0)}(t_2) = 0$ from the initial condition. Equation (1.90) shows that $c_n^{(0)}(t_2) = c_n^{(0)}(t_1)$, so $c_f^{(0)}(t_2) = 0$. We find from Equation (1.91)

$$\lambda \dot{c}_f^{(1)}(t) = -i\langle f|\lambda H_1|i\rangle \mathrm{e}^{i(E_f - E_i)t} \tag{1.94}$$

$$= -i\langle f|\lambda V|i\rangle \mathrm{e}^{i(E_f - E_i \pm \omega)t} \tag{1.95}$$

We integrate this equation and obtain

$$\int_{t_1}^{t_2} \lambda \dot{c}_f^{(1)} \, dt = \lambda[c_f^{(1)}(t_2) - c_f^{(1)}(t_1)] \tag{1.96}$$

$$= -i\langle f|\lambda V|i\rangle \int_{t_1}^{t_2} dt \, \mathrm{e}^{i(E_f - E_i \pm \omega)t} \tag{1.97}$$

We know from the initial condition that $c_f^{(1)}(t_1) = 0$. We let $t_1 \to -\infty$ and $t_2 \to +\infty$ and use

$$\frac{1}{2\pi} \int_{-\infty}^{\infty} dt \, \mathrm{e}^{i\omega t} = \delta(\omega) \tag{1.98}$$

to get

$$\lambda c_f^{(1)}(\infty) = 2\pi i \langle f|\lambda V|i\rangle \delta(E_f - E_i \pm \omega) \tag{1.99}$$

We see that the amplitude for the system to be in the state $|f\rangle$ at $t = \infty$ is given by $c_f(\infty) = c_f^{(1)}(\infty)$ to first order in V, and is non-zero if $\langle f|\lambda V|i\rangle \neq 0$ and $E_f = E_i \mp \hbar\omega$. This corresponds to a selection rule and a change in the energy of the system by $\mp\omega$. The meaning of this will become clear soon.

We obtain the transition probability per unit time (or 'transition probability' for short) w_{fi} from Equation (1.99) as

$$dw_{fi} = \lim_{T\to\infty} \frac{|c_f(\infty)|^2}{T} = \lim_{T\to\infty} \frac{|\lambda c_f^{(1)}(\infty)|^2}{T}$$

$$= \lim_{T\to\infty} \frac{4\pi^2}{T} \left|\langle f|\lambda V|i\rangle\right|^2 [\delta(E_f - E_i \pm \omega)]^2 \tag{1.100}$$

with $T = t_2 - t_1$. We used that $c_f^{(0)}(\infty) = 0$; see the discussion just above Equation (1.94). The δ-function squared is evaluated as

$$[\delta(E)]^2 = \delta(E) \lim_{T\to\infty} \frac{1}{2\pi} \int_{-T/2}^{+T/2} dt \, \mathrm{e}^{iEt} \tag{1.101}$$

$$= \delta(E) \lim_{T\to\infty} \frac{T}{2\pi} \tag{1.102}$$

So finally we have

$$dw_{fi} = 2\pi \, |\langle f|\lambda V|i\rangle|^2 \, \delta(E_f - E_i \pm \omega) \tag{1.103}$$

This result is to first order in V and is called Fermi's Golden Rule. For some applications the first order calculation is not precise enough, especially if it is zero, and we must go to second order. We will leave that calculation until we need its result in the subsection on photon scattering.

The transition probability (per unit time) is related to the lifetime τ of a state as

$$\tau = \frac{1}{w_{fi}} \tag{1.104}$$

This can be seen as follows. If a system has a probability $P_i(t)$ to be in the state $|i\rangle$ at time t, then a time dt later that probability will have changed by $dP_i = -w_{fi}P_i dt$. We can integrate this relation to get $P_i(t) = P_i(0)\,\mathrm{e}^{-w_{fi}t}$. Using the definition of lifetime τ from $P_i(t) = P_i(0)\,\mathrm{e}^{-t/\tau}$ we find Equation (1.104).

In using Equation (1.77) and Equation (1.78) we see that H_0 consists of two parts, one for a charged particle in an electrostatic electric field (the first two terms) and one for the electromagnetic field (the third term). These two parts in H_0 are independent of each other. We assume that we know the eigenkets of each part so we can form the eigenkets of H_0 as product eigenkets as appropriate for a Hamiltonian that consists of two independent parts. The product eigenkets are formed from the eigenkets $|A\rangle, |B\rangle, \cdots$ of the charged particle part of H_0 and the eigenkets $|\mathbf{k}, \lambda\rangle$ of the electromagnetic part of H_0. The eigenvalues of H_0 are the sum of the eigenvalues of the two parts. Consider as an example a system that decays from state A to state B under photon emission

$$A \to B + \gamma \tag{1.105}$$

The initial state is represented by $|A\rangle|0\rangle$ and $|B\rangle|\mathbf{k}, \lambda\rangle$ or more compactly $|A; 0\rangle$ and the final state by $|B; \mathbf{k}, \lambda\rangle$.

We will now apply this formalism to the case of spontaneous emission of a photon. This was first discussed by Einstein in 1917 (having finished his theory of General Relativity) when he introduced his A and B coefficients that describe spontaneous and induced emission respectively. Examples of spontaneous emission can be found in atomic, molecular and nuclear physics. It plays an important role in black body radiation.

1.4 SPONTANEOUS EMISSION

1.4.1 First Order Result

We consider the process

$$A \to B + \gamma \tag{1.106}$$

and use the Hamiltonian Equation (1.77). The Harmonic Perturbation is given by the last two terms

$$\lambda H_1 = -\frac{e}{m}\mathbf{A}\cdot\mathbf{p} + \frac{e^2}{2m}\mathbf{A}^2 \tag{1.107}$$

with $\mathbf{A}$ given by Equation (1.41)

$$\mathbf{A}(x) = \sqrt{\frac{4\pi}{V}}\sum_{\mathbf{k},\lambda}\frac{1}{\sqrt{2\omega_k}}[a_\lambda(\mathbf{k})\,e^{ikx}\boldsymbol{\varepsilon}_\lambda(\mathbf{k}) + a_\lambda^\dagger(\mathbf{k})\,e^{-ikx}\boldsymbol{\varepsilon}_\lambda^*(\mathbf{k})] \tag{1.108}$$

which contains harmonic factors of the form $\exp(\pm i\omega t)$, just as we considered in the previous section. We use Equation (1.94) to get

$$\begin{aligned}\lambda\,\dot{c}_f^{(1)} &= -i\langle f|\lambda H_1|i\rangle e^{i(E_f-E_i)t}\\ &= -i\langle B;\mathbf{k},\lambda| -\frac{e}{m}\mathbf{A}\cdot\mathbf{p} + \frac{e^2}{2m}\mathbf{A}^2|A;0\rangle e^{i(E_B-E_A)t}\end{aligned} \tag{1.109}$$

We assigned zero energy to the electromagnetic term in H_0 so $E_f = E_B$. The term linear in $\mathbf{A}$ creates or annihilates a photon through $a_\lambda^\dagger(\mathbf{k})$ and $a_\lambda(\mathbf{k})$ so it changes the number of photons by ± 1. It can be seen that the $\mathbf{A}^2$ term changes the number of photons by 0 or ± 2 so this term does not contribute to Equation (1.106) because the product with the bra $\langle\mathbf{k},\lambda|$ would be zero. In the present case we must create one photon so the product with $\langle\mathbf{k},\lambda|$ must be non-zero. So only the $a_\lambda^\dagger(\mathbf{k})$ term in $\mathbf{p}\cdot\mathbf{A}$ contributes. We obtain

$$\lambda\,\dot{c}_f^{(1)} = -i\sqrt{\frac{4\pi}{V2\omega}}\left(-\frac{e}{m}\right)\langle B;\mathbf{k},\lambda|a_\lambda^\dagger(\mathbf{k})e^{-i\mathbf{k}\cdot\mathbf{x}}\,\boldsymbol{\varepsilon}_\lambda^*(\mathbf{k})\cdot\mathbf{p}|A;0\rangle e^{i(E_B-E_A+\omega)t} \tag{1.110}$$

where $\mathbf{x}$ is the coordinate of the charged particle (for example the electron in a H-atom) and $\mathbf{p}$ is its momentum, while $\mathbf{k}$ is the momentum of the photon. We know that $a_\lambda^\dagger(\mathbf{k})\,|0\rangle = \sqrt{1}\,|\mathbf{k},\lambda\rangle$ (the creation operator in action) so $\langle\mathbf{k},\lambda|a_\lambda^\dagger(\mathbf{k})|0\rangle = 1$ and one bra and one ket disappears from the product bra and ket. Thus we get

$$\lambda\,\dot{c}_f^{(1)} = \frac{ie}{m}\sqrt{\frac{4\pi}{V2\omega_k}}\langle B|e^{-i\mathbf{k}\cdot\mathbf{x}}\,\boldsymbol{\varepsilon}_\lambda^*(\mathbf{k})\cdot\mathbf{p}|A\rangle e^{i(E_B-E_A+\omega)t} \tag{1.111}$$

If we compare this expression with the general expression in Equation (1.95) for a harmonic perturbation we find that

$$\lambda V = \frac{ie}{m}\sqrt{\frac{4\pi}{V2\omega_k}}e^{-i\mathbf{k}\cdot\mathbf{x}}\,\boldsymbol{\varepsilon}_\lambda^*(\mathbf{k})\cdot\mathbf{p} \tag{1.112}$$

Substitution of this expression in Fermi's Golden Rule Equation (1.103) gives

$$dw_{fi} = 2\pi \left|\langle B|\lambda V|A\rangle\right|^2 \delta(E_b - E_A \pm \omega)$$
$$= 2\pi \left(\frac{e}{m}\right)^2 \frac{4\pi}{V2\omega_k} \left|\langle B|e^{-i\mathbf{k}\cdot\mathbf{x}}\,\boldsymbol{\varepsilon}^*_\lambda(\mathbf{k})\cdot\mathbf{p}|A\rangle\right|^2 \delta(E_B - E_A + \omega) \quad (1.113)$$

Note that the transition probability is of $O(e^2)$, that the δ-function requires that $E_A = E_B + \omega$ and that the operator V must be such that $\langle B|V|A\rangle \neq 0$. The expression in Equation (1.113) is rather complicated because of the exponential. We can expand the exponential in a Taylor series and see what we can learn, at least from the first few terms.

1.4.2 Dipole Transition

Keeping for the moment only the first term of the expansion of the exponential in Equation (1.113) we get

$$dw_{fi} = 2\pi \left(\frac{e}{m}\right)^2 \frac{4\pi}{V2\omega_k} \left|\langle B|\boldsymbol{\varepsilon}^*_\lambda(\mathbf{k})\cdot\mathbf{p}|A\rangle\right|^2 \delta(E_B - E_A + \omega) \qquad (1.114)$$

This is called the 'Dipole Approximation' for reasons that will become clear shortly. The matrix element $\langle B|\mathbf{p}|A\rangle$ can be evaluated as follows

$$\langle B|\mathbf{p}|A\rangle = m\langle B|\dot{\mathbf{x}}|A\rangle$$
$$= im\langle B|[H_0, \mathbf{x}]|A\rangle$$
$$= im\,(E_B - E_A)\langle B|\mathbf{x}|A\rangle \qquad (1.115)$$

In the first step we used the non-relativistic relation between momentum and velocity, in the second we used

$$\dot{\mathbf{x}} = \frac{d\mathbf{x}}{dt} = i\,[H, \mathbf{x}] \qquad (1.116)$$

with H equal to the charged particle piece of Equation (1.78). The matrix element $\langle B|\mathbf{x}|A\rangle$ can be calculated because we assumed we know the eigenkets $|A\rangle$ and $|B\rangle$. The answer depends of course on the specific problem and its eigenkets. For now we define a quantity, the (transition) Dipole Moment, as

$$\mathbf{D} = \langle B|e\mathbf{x}|A\rangle \qquad (1.117)$$

and use it with Equation (1.115) in Equation (1.114) to get

$$dw_{fi} = \frac{4\pi^2}{V}\,\omega_k\,|\,\boldsymbol{\varepsilon}^*_\lambda(\mathbf{k})\cdot\mathbf{D}\,|^2\,\delta(E_B - E_A + \omega_k) \qquad (1.118)$$

We have used the constraint imposed by the delta function when we replaced $(E_B - E_A)^2$ by ω_k^2. We now see why this is called the Dipole Approximation: the transition probability depends upon the dipole moment (squared, as in classical radiation theory). The factor $\boldsymbol{\varepsilon}_\lambda^*(\mathbf{k}) \cdot \mathbf{D}$ implies that the probability for photon emission is a maximum when $\boldsymbol{\varepsilon}_\lambda^*(\mathbf{k})$ is parallel to $\mathbf{D}$ and thus $\mathbf{k}$ perpendicular to $\mathbf{D}$, again in agreement with classical radiation theory for dipole radiation. We also see that energy is conserved because the energy $E_B + \omega_k$ in the final state is equal to E_A, the energy in the initial state, thanks to the δ-function. Look over the calculation to see where the $+$ sign comes from and verify that photon absorption would involve $a_\lambda(\mathbf{k})$ and give $\delta(E_B - E_A - \omega_k)$ (note the $-$ sign in front of ω_k) automatically and the transition probability would again conserve energy. As mentioned before, $\mathbf{D}$ is calculable for specific cases but one can sometimes see from symmetry that $\mathbf{D}$ has to be zero. An example is radiation in a H-atom when it transitions from a $2s$ state to a $1s$ state. The s states have $l = 0$ and are therefore spherically symmetric, $e\mathbf{x}$ is odd, therefore the transition dipole moment is zero. Another way to see this is by looking at the integration over space to be done to calculate $\mathbf{D}$. The wave functions are even functions but $\mathbf{x}$ is odd so the integral is zero.

Experimentally we ask for the probability that a photon is emitted in a specific region of phase space $d^3\mathbf{k}$. We know from statistical physics that such a phase space volume is given by

$$\frac{Vd^3\mathbf{k}}{(2\pi)^3} = \frac{V\mathbf{k}^2 d|\mathbf{k}|\, d\Omega_k}{(2\pi)^3} = \frac{V\omega_k^2\, d\omega_k\, d\Omega_k}{(2\pi)^3} \tag{1.119}$$

The product of the transition probability w_{fi} and the volume of phase space in which we find the photon is the desired quantity

$$dw_{fi} = \frac{4\pi^2}{V}\omega_k \,|\, \boldsymbol{\varepsilon}_\lambda^*(\mathbf{k}) \cdot \mathbf{D} \,|^2\, \frac{V\omega_k^2 d\omega_k d\Omega_k}{(2\pi)^3}\,\delta(E_B - E_A + \omega_k) \tag{1.120}$$

We see that V cancels as it should (so we can let $V \to \infty$). We can integrate over the photon energy ω_k because it is constrained by the δ-function, enforcing energy conservation. If the experiment does not observe the polarization of the photon we must sum over λ. We get

$$\frac{dw_{fi}}{d\Omega_k} = \frac{\omega_k^3}{2\pi} \sum_{\lambda=1,2} |\, \boldsymbol{\varepsilon}_\lambda^*(\mathbf{k}) \cdot \mathbf{D} \,|^2 \tag{1.121}$$

We perform the sum over λ using the three vectors $\boldsymbol{\varepsilon}_1(\mathbf{k}), \boldsymbol{\varepsilon}_2(\mathbf{k}), \mathbf{k}/\omega_k$ that form a triplet of (real) unit vectors. We can write

$$|\mathbf{D}|^2 = D_x^2 + D_y^2 + D_z^2 = (\boldsymbol{\varepsilon}_1 \cdot \mathbf{D})^2 + (\boldsymbol{\varepsilon}_2 \cdot \mathbf{D})^2 + \left(\frac{\mathbf{k} \cdot \mathbf{D}}{\omega_k}\right)^2 \tag{1.122}$$

Define the angle θ between $\mathbf{k}$ and $\mathbf{D}$ by $\mathbf{k}\cdot\mathbf{D} = \omega_k\,|\mathbf{D}|\,\cos\theta$. Then

$$\sum_{\lambda=1,2} (\boldsymbol{\varepsilon}_\lambda(\mathbf{k})\cdot\mathbf{D})^2 = (\boldsymbol{\varepsilon}_1\cdot\mathbf{D})^2 + (\boldsymbol{\varepsilon}_2\cdot\mathbf{D})^2$$
$$= |\mathbf{D}|^2\,(1-\cos^2\theta) = \sin^2\theta\,|\mathbf{D}|^2 \qquad (1.123)$$

We see that the probability is a maximum if $\theta = \pi/2$ or $\mathbf{k}\perp\mathbf{D}$ as noted earlier. The integral over θ is trivial and we get for the total transition probability

$$w_{fi} = \frac{4}{3}\,\omega_k^3\,|\mathbf{D}|^2 \qquad (1.124)$$

The luminosity $\mathcal{L}$ of the emitted radiation is obtained by multiplying w_{fi} by the energy ω of each photon so we get

$$\mathcal{L} = \frac{4}{3}\omega^4\,|\mathbf{D}|^2 \qquad (1.125)$$

As an example we calculate the probability for spontaneous emission of a photon by a H-atom in the transition $2p \to 1s$. We have that $E_n = -(1/2)\,m\alpha^2/n^2$ with m the mass of the electron, α the fine-structure constant, and n the principal quantum number. For the $2p \to 1s$ transition we find that $\omega_k = (3/8)\,m\alpha^2$. We guess that the dipole moment is equal to the electron's charge times the Bohr radius $a_0 = 1/(\alpha m)$. Altogether we get $w_{fi} = (9/128)\,m\alpha^5$, not very large because $\alpha^2 \approx 1/137$. To get a numerical value we must insert the correct powers of $\hbar$ and c using dimensional analysis. We find that we need to insert a factor $c^2/\hbar$ to get $w_{fi} = (9/128)\,m\alpha^5(c^2/\hbar) = 1.1\times 10^9\,\mathrm{s}^{-1}$. The lifetime τ of a state is the inverse of the transition probability and we find that $\tau = 0.9\,\mathrm{ns}$, a very long time indeed (on an atomic scale). With the factor $c^2/\hbar$ in Equation (1.124) and c^2 in Equation (1.125), we see that the luminosity $\mathcal{L}$ does not contain $\hbar$. This allows for the classical calculation of the luminosity to be done with a result in agreement with Equation (1.125).

1.4.3 Higher Multipole Transition

We now consider the next (second) term in the Taylor series' expansion of the exponential in Equation (1.113). The matrix element of that second term is

$$\langle B;\mathbf{k},\lambda|(\mathbf{k}\cdot\mathbf{x})(\boldsymbol{\varepsilon}^*\cdot\mathbf{p})|A\rangle \qquad (1.126)$$

The original order of the position coordinate $\mathbf{x}$ and the momentum operator $\mathbf{p}$ is kept as it was in the full matrix element Equation (1.113) because $\mathbf{p}$ acts of course on $\mathbf{x}$. We can manipulate the operator $(\mathbf{k}\cdot\mathbf{x})(\boldsymbol{\varepsilon}^*\cdot\mathbf{p})$ into a form that consists of two terms, one representing a magnetic moment, the other an electric quadrupole moment

$$(\mathbf{k}\cdot\mathbf{x})(\boldsymbol{\varepsilon}^*\cdot\mathbf{p}) = \frac{1}{2}[(\mathbf{k}\cdot\mathbf{x})(\boldsymbol{\varepsilon}^*\cdot\mathbf{p}) - (\boldsymbol{\varepsilon}^*\cdot\mathbf{x})(\mathbf{k}\cdot\mathbf{p})] + \frac{1}{2}[(\mathbf{k}\cdot\mathbf{x})(\boldsymbol{\varepsilon}^*\cdot\mathbf{p}) + (\boldsymbol{\varepsilon}^*\cdot\mathbf{x})(\mathbf{k}\cdot\mathbf{p})] \tag{1.127}$$

where we have subtracted and added the term $(\boldsymbol{\varepsilon}^*\cdot\mathbf{x})(\mathbf{k}\cdot\mathbf{p})$. Now consider the expression $(\mathbf{k}\times\boldsymbol{\varepsilon}^*)\cdot(\mathbf{x}\times\mathbf{p})$ which we will cast in another form while keeping the order of $\mathbf{x}$ and $\mathbf{p}$ the same

$$\begin{aligned}(\mathbf{k}\times\boldsymbol{\varepsilon}^*)\cdot(\mathbf{x}\times\mathbf{p}) &= \mathbf{x}\cdot[\mathbf{p}\times(\mathbf{k}\times\boldsymbol{\varepsilon}^*)]\\ &= \mathbf{x}\cdot[(\mathbf{p}\cdot\boldsymbol{\varepsilon}^*)\mathbf{k} - (\mathbf{p}\cdot\mathbf{k})\boldsymbol{\varepsilon}^*]\\ &= (\mathbf{x}\cdot\mathbf{k})(\mathbf{p}\cdot\boldsymbol{\varepsilon}^*) - (\mathbf{x}\cdot\boldsymbol{\varepsilon}^*)(\mathbf{p}\cdot\mathbf{k})\end{aligned} \tag{1.128}$$

In the first line we can identify $\mathbf{k}\times\boldsymbol{\varepsilon}^*$ with the magnetic field $\mathbf{B}$ associated with the photon according to Equation (1.45) and $\mathbf{x}\times\mathbf{p}$ with the angular momentum $\mathbf{L}$ of the charged particle. The first equality is obtained by a left-cyclic permutation maintaining the order of $\mathbf{x}$ and $\mathbf{p}$. The term in the last line is twice the term in the first line of Equation (1.128). We know from the discussion of an atom in a static external magnetic field that the orbital angular momentum $\mathbf{L}$ is associated with a magnetic moment $e/(2m)\mathbf{L}$. So we define a (transition) magnetic moment

$$\langle B|\frac{e}{2m}\mathbf{L}|A\rangle = \boldsymbol{\mu} \tag{1.129}$$

so that we interpret the first line in the matrix element in Equation (1.128) as a transition due to the interaction $\boldsymbol{\mu}\cdot\mathbf{B}$ of the magnetic dipole moment of the charged particle with the magnetic field associated with the photon. This is in close analogy with the electric dipole transition encountered earlier, except that the magnetic dipole transition probability is much smaller.

The first term in the last line of Equation (1.128) can be written

$$(\mathbf{k}\cdot\mathbf{x})(\boldsymbol{\varepsilon}^*\cdot\mathbf{p}) = (\mathbf{k}\cdot\mathbf{x})\,\{\boldsymbol{\varepsilon}^*\cdot[H_0,\mathbf{x}]\,im\} \tag{1.130}$$

where $\mathbf{p}$ is dealt with in a way similar to the manipulation in Equation (1.115). For the second term in the last line of Equation (1.128) we use that $(\boldsymbol{\varepsilon}^*\cdot\mathbf{x})(\mathbf{k}\cdot\mathbf{p}) = (\mathbf{k}\cdot\mathbf{p})(\boldsymbol{\varepsilon}^*\cdot\mathbf{x})$, in other words we can change the order of $\mathbf{p} \cdot \mathbf{k}$ and $\boldsymbol{\varepsilon}^* \cdot \mathbf{x}$. That this is so can be seen as follows ($|\psi\rangle$ is an arbitrary ket)

$$\begin{aligned}
(\mathbf{k} \cdot \mathbf{p})(\boldsymbol{\varepsilon}^* \cdot \mathbf{x})|\psi\rangle &= p_i k_i \varepsilon_j^* x_j \,|\psi\rangle \\
&= \frac{1}{i}\frac{\partial}{\partial x_i} k_i \varepsilon_j^* x_j \,|\psi\rangle \\
&= \frac{1}{i} k_i \varepsilon_j^* \delta_{ij} \,|\psi\rangle + \frac{1}{i} k_i \varepsilon_j^* x_j \frac{\partial}{\partial x_i} |\psi\rangle \\
&= \frac{1}{i}(\boldsymbol{\varepsilon}^* \cdot \mathbf{k})\,|\psi\rangle + k_i \varepsilon_j^* x_j p_i \,|\psi\rangle \\
&= (\boldsymbol{\varepsilon}^* \cdot \mathbf{x})(\mathbf{k} \cdot \mathbf{p})|\psi\rangle
\end{aligned} \tag{1.131}$$

The term with $\boldsymbol{\varepsilon}_\lambda^*(\mathbf{k})$ is zero in the Coulomb Gauge. The second term in the second line of Equation (1.128) becomes with Equation (1.131)

$$(\boldsymbol{\varepsilon}^* \cdot \mathbf{x})(\mathbf{k} \cdot \mathbf{p}) = \{\mathbf{k} \cdot [H_0, \mathbf{x}]\, im\}\,(\boldsymbol{\varepsilon}^* \cdot \mathbf{x}) \tag{1.132}$$

We must add the relations Equation (1.130) and Equation (1.132). As we do this we keep k to the left

$$k_j x_j \varepsilon_i [H_0, x_i]\, im + k_j [H_0, x_j]\, \varepsilon_i x_i\, im = \varepsilon_i k_j [H_0, x_i x_j]\, im \tag{1.133}$$

To get the corresponding matrix element we must sandwich this result between $\langle A|$ and $|B\rangle$

$$\langle B|[H_0, x_i x_j]|A\rangle = (E_B - E_A)\langle B|x_i x_j|A\rangle = \omega_k \frac{Q_{ij}}{e} \tag{1.134}$$

where the electric quadrupole moment Q_{ij} is

$$Q_{ij} = e\langle B|x_i x_j - \frac{1}{3}\delta_{ij}\mathbf{x}^2|A\rangle \tag{1.135}$$

The extra term proportional to δ_{ij} in Equation (1.135) does not matter because in Equation (1.133) it gets multiplied by $\varepsilon_i k_j$ giving $\boldsymbol{\varepsilon} \cdot \mathbf{k}$ which is zero in the Coulomb Gauge, see Equation (1.19). We interpret this as an electric quadrupole transition. Its transition probability is of the same order of magnitude as the one from the magnetic dipole moment and much smaller then the transition probability from the electric dipole moment.

The various multipole transitions are called $E1$ for electric dipole, M1 for magnetic dipole, E2 for electric quadrupole, M2 for magnetic quadrupole, etc., transitions. The order of magnitudes of their transition probabilities are respectively $\alpha(\text{size}/\lambda)^2$ for $E1$, $\alpha(\text{size}/\lambda)^4$ for $M1$ and $E2$, ... where λ is the wavelength of the emitted photon and α is the fine structure constant. For the $E1$ transitions this can be seen from Equation (1.124): one power of ω_k provides the unit ($\sec^{-1}$) for w_{fi}, the other two powers of ω_k are proportional to $1/\lambda^2$. The dipole moment is of order of the electron charge times the (linear) size of the system and their product enters squared in the expression in Equation (1.124). The $M1$ and $E2$ multipole transitions result from the next term in the Taylor expansion of $\exp(-\mathbf{k}\cdot\mathbf{x})$ in Equation (1.113). This term has a factor $\mathbf{k}\cdot\mathbf{x}$ more in the amplitude (that factor squared in their probabilities), and that factor is of order size/λ. The extension of these considerations to even higher multipoles is obvious. We summarize these considerations in Table 1.2. The parity selection rule will be discussed in Section 1.6 and is listed here for completeness.

In atomic physics it is usually sufficient to consider only the lowest non-zero multipole in calculating transition probabilities. The situation is different in nuclear transitions because typical energy releases in a nucleus are of order of MeVs, a factor 10^5 larger than in atomic transitions, leading to wavelengths that are a factor 10^5 smaller. Furthermore the nucleus is about a factor 10^4 smaller then an atom. Therefore the ratio (size/λ) is an order of magnitude larger in nuclear transitions than in atomic transitions. This enhances the relative magnitude of higher multipole transitions in nuclear transitions, and these present therefore excellent opportunities to observe higher order multipole transitions. Each type of multipole transition obeys their own selection rules for their transition probabilities. We will return to these in the next subsection.

We note that there is no dipole radiation in gravitational radiation because the dipole moment of a mass distribution is identically zero by definition of its center-of-mass. Thus the lowest order multipole that can produce gravitational radiation is the quadrupole moment of the radiating mass distribution. This requires a non-spherical mass distribution, making the detection of gravitational radiation even harder to observe than it already is because of the weakness of the gravitational interaction relative to the electromagnetic interaction.

It is clear that there must be a systematic development from dipole moments to higher multipole moments. This can indeed be achieved when using vector spherical harmonics. This is outside the scope of this book.

Table 1.2 Parity selection rules and order of magnitude of w_{fi}

Transition	Parity Change	Magnitude of w_{fi}
$E1$	Yes	$\alpha\,(\text{size}/\lambda)^2$
$M1$	No	$\alpha\,(\text{size}/\lambda)^4$
$E2$	Yes	$\alpha\,(\text{size}/\lambda)^4$
$M2$	No	$\alpha\,(\text{size}/\lambda)^6$
$E3$	Yes	$\alpha\,(\text{size}/\lambda)^6$
..	..	..

1.5 BLACKBODY RADIATION

Blackbody radiation played a most crucial role in the development of quantum physics through the work of Planck, who introduced in 1901 the notion of the quantization of electromagnetic energy leading to a correct description of the energy spectrum of blackbody radiation. In 1917, well before the development of Quantum Physics, Einstein revisited blackbody radiation when he introduced his well known A and B coefficients and re-derived Planck's energy spectrum of blackbody radiation. The Einstein coefficient A describes spontaneous emission, while B (two coefficients really) describes induced emission and induced absorption of radiation. Einstein considered a quantum physical system that has two states, a ground state and an excited state. The Einstein coefficient A is the probability that the system makes a 'spontaneous' transition from the excited state to the ground state with the emission of a photon. The Einstein coefficient $B_{1\to 2}$ is the probability that the system goes from the ground state to the excited state under absorption of a photon and $B_{2\to 1}$ to go from the excited state to the ground state under absorption of a photon and the subsequent emission of two photons. Einstein was able to show using a thermodynamic argument that the two B coefficients are equal.

In view of its importance we follow in Einstein's footsteps, but treat blackbody radiation with the tools of second quantization, an option that Einstein did not have at the time. We consider a quantum physical system that has two states, a ground state $|1\rangle$ with energy E_1 and an excited state $|2\rangle$ with energy E_2. In the presence of electromagnetic radiation we will have transitions

$$\gamma + 1 \leftrightarrow 2 \tag{1.136}$$

If an assembly of identical systems is in equilibrium with radiation we have from the Boltzmann distribution that

$$\frac{n_1}{n_2} = \frac{\mathrm{e}^{-E_1/kT}}{\mathrm{e}^{-E_2/kT}} = \mathrm{e}^{\hbar\omega/kT} \geq 1 \tag{1.137}$$

where n_1 and n_2 are the number of systems in state $|1\rangle$ and $|2\rangle$ respectively and $E_2 - E_1 = \hbar\omega$. If we have initially $n_\lambda(\mathbf{k})$ photons, then the matrix element for absorption of one of the photons is

$$\langle 2; n_\lambda(\mathbf{k}) - 1 | a_\lambda(\mathbf{k})\, \mathrm{e}^{i\mathbf{k}\cdot\mathbf{x}} \boldsymbol{\varepsilon}_\lambda(\mathbf{k}) \cdot \mathbf{p} | 1; n_\lambda(\mathbf{k}) \rangle = \sqrt{n_\lambda(\mathbf{k})}\, \langle 2 | \mathrm{e}^{i\mathbf{k}\cdot\mathbf{x}} \boldsymbol{\varepsilon}_\lambda(\mathbf{k}) \cdot \mathbf{p} | 1 \rangle \tag{1.138}$$

We must annihilate a photon with momentum $\mathbf{k}$ and polarization λ from the initial state and select therefore the operator $a_\lambda(\mathbf{k})$. This operator is accompanied by the exponential with a $+$ sign in its exponent and $\boldsymbol{\varepsilon}_\lambda(\mathbf{k})$.

Equation (1.34) has been used to obtain the second line in Equation (1.138). We do not make the 'Dipole Approximation' but keep the expression completely general, except that we limit ourselves to terms of order e in the perturbation expansion. Note that Equation (1.138) 'explains' why it is that radiation is absorbed one photon at a time, a fact that had to be postulated in the early development of quantum physics. The matrix element for emission of a photon is

$$\langle 1; n_\lambda(\mathbf{k}) + 1 | a_\lambda^\dagger(\mathbf{k})\, \mathrm{e}^{-i\mathbf{k}\cdot\mathbf{x}} \boldsymbol{\varepsilon}_\lambda^*(\mathbf{k}) \cdot \mathbf{p} | 2; n_\lambda(\mathbf{k}) \rangle = \sqrt{n_\lambda(\mathbf{k}) + 1}\, \langle 1 | \mathrm{e}^{-i\mathbf{k}\cdot\mathbf{x}} \boldsymbol{\varepsilon}_\lambda^*(\mathbf{k}) \cdot \mathbf{p} | 2 \rangle \tag{1.139}$$

This equation 'explains' that radiation is emitted one photon at a time. We have left out factors that are common to the matrix elements in Equation (1.138) and Equation (1.139). To calculate transition probabilities for Fermi's Golden Rule, we need to square the matrix elements of Equation (1.138) and Equation (1.139). The probability for photon absorption is thus proportional to $n_\lambda(\mathbf{k})$ and the probability for photon emission is proportional to $n_\lambda(\mathbf{k}) + 1$. If there are no photons present initially ($n_\lambda(\mathbf{k}) = 0$) then the probability for photon absorption is zero, but the probability for photon emission is not. Because induced absorption and induced emission require by definition a non-zero number of photons in the initial state, we identify terms in the squared amplitudes that are proportional to $n_\lambda(\mathbf{k})$ with induced transitions and the leftover term in the square of the amplitude in Equation (1.139) with spontaneous emission, because the latter is non-zero even if the number of photons in the initial state $n_\lambda(\mathbf{k}) = 0$. Thus we identify the probability for induced absorption of a photon with

$$B_{1\to 2} \propto |\langle 2 | \mathrm{e}^{i\mathbf{k}\cdot\mathbf{x}} \boldsymbol{\varepsilon}_\lambda(\mathbf{k}) \cdot \mathbf{p} | 1 \rangle|^2 \tag{1.140}$$

and identify creation of a photon with

$$B_{2\to 1} \propto |\langle 1 | \mathrm{e}^{-i\mathbf{k}\cdot\mathbf{x}} \boldsymbol{\varepsilon}_\lambda^*(\mathbf{k}) \cdot \mathbf{p} | 2 \rangle|^2 \tag{1.141}$$

where common factors have been omitted. The B are the Einstein B coefficients. We can show that $B_{12} = B_{21}$ as follows

$$B_{12} = \langle 2 | \mathrm{e}^{i\mathbf{k}\cdot\mathbf{x}} \boldsymbol{\varepsilon}_\lambda(\mathbf{k}) \cdot \mathbf{p} | 1 \rangle^* \langle 2 | \mathrm{e}^{i\mathbf{k}\cdot\mathbf{x}} \boldsymbol{\varepsilon}_\lambda(\mathbf{k}) \cdot \mathbf{p} | 1 \rangle \tag{1.142}$$

$$= \langle 1 | \boldsymbol{\varepsilon}_\lambda^*(\mathbf{k}) \cdot \mathbf{p}\, \mathrm{e}^{-i\mathbf{k}\cdot\mathbf{x}} | 2 \rangle \langle 1 | \boldsymbol{\varepsilon}_\lambda^*(\mathbf{k}) \cdot \mathbf{p}\, \mathrm{e}^{-i\mathbf{k}\cdot\mathbf{x}} | 2 \rangle^* \tag{1.143}$$

$$= \langle 1 | \mathrm{e}^{-i\mathbf{k}\cdot\mathbf{x}}\, \boldsymbol{\varepsilon}_\lambda^*(\mathbf{k}) \cdot \mathbf{p} | 2 \rangle \langle 1 | \mathrm{e}^{-i\mathbf{k}\cdot\mathbf{x}}\, \boldsymbol{\varepsilon}_\lambda^*(\mathbf{k}) \cdot \mathbf{p} | 2 \rangle^* \tag{1.144}$$

$$= B_{21} \tag{1.145}$$

The second line Equation (1.143) is obtained from Equation (1.142) by taking the complex conjugate of Equation (1.142) and using the general relation $\langle\psi|ABC|\phi\rangle^* = \langle\phi|C^\dagger B^\dagger A^\dagger|\psi\rangle$ twice. To obtain Equation (1.144) we moved the factor $\boldsymbol{\varepsilon}^*_\lambda(\mathbf{k})\cdot\mathbf{p}$ through the exponential $\mathrm{e}^{-i\mathbf{k}\cdot\mathbf{x}}$. This is allowed because in the Coulomb Gauge $\mathbf{k}\cdot\boldsymbol{\varepsilon}_\lambda(\mathbf{k}) = 0$, see Equation (1.19), and thus we have with the test function ψ

$$\begin{aligned}\boldsymbol{\varepsilon}^*_\lambda(\mathbf{k})\cdot\mathbf{p}\,\mathrm{e}^{-i\mathbf{k}\cdot\mathbf{x}}\,\psi &= \frac{1}{i}\,\boldsymbol{\varepsilon}^*_\lambda(\mathbf{k})\cdot\nabla\mathrm{e}^{-i\mathbf{k}\cdot\mathbf{x}}\psi \\ &= \frac{1}{i}\,\boldsymbol{\varepsilon}^*_\lambda(\mathbf{k})\cdot(-i\mathbf{k})\,\mathrm{e}^{-i\mathbf{k}\cdot\mathbf{x}}\,\psi + \frac{\mathrm{e}^{-i\mathbf{k}\cdot\mathbf{x}}}{i}\,\boldsymbol{\varepsilon}^*_\lambda(\mathbf{k})\cdot\nabla\psi \\ &= \mathrm{e}^{-i\mathbf{k}\cdot\mathbf{x}}\,\boldsymbol{\varepsilon}^*_\lambda(\mathbf{k})\cdot\mathbf{p}\,\psi \end{aligned} \tag{1.146}$$

We find that $B_{12} = B_{21}(= B)$ as Einstein proved in a different manner. He considered the limiting case of an infinite number of photons at $T = \infty$ in equilibrium with an ensemble of identical two-state systems. Induced transitions dominate the spontaneous ones and $n_1 = n_2$ at $T = \infty$ according to Equation (1.137). In equilibrium n_1 and n_2 are constant and therefore the number of systems going from $1 \to 2$ and from $2 \to 1$ must be equal, so

$$n_1 B_{1\to2} = n_2 B_{2\to1} \tag{1.147}$$

With $n_1 = n_2$ we find that $B_{1\to2} = B_{2\to1}$ and this must be true at all T because the Einstein coefficients do not depend upon T. The Einstein coefficient A has been calculated in the previous section on spontaneous emission.

We complete Einstein's derivation of the energy spectrum of blackbody radiation at finite temperature. Equilibrium at any temperature requires

$$n_1 w_{1\to2} = n_2 w_{2\to1} \tag{1.148}$$

Using this relation and Equation (1.137) we get

$$\frac{n_1}{n_2} = \frac{w_{2\to1}}{w_{1\to2}} = \mathrm{e}^{\hbar\omega/kT} \tag{1.149}$$

We also have that

$$\frac{w_{2\to1}}{w_{1\to2}} = \frac{\text{prob for emission}}{\text{prob for absorption}} = \frac{n_\lambda(\mathbf{k})+1}{n_\lambda(\mathbf{k})} \tag{1.150}$$

If we combine Equation (1.149) and Equation (1.150) we can solve for $n_\lambda(\mathbf{k})$ and get

$$n_\lambda(\mathbf{k}) = \frac{1}{\mathrm{e}^{\hbar\omega/kT} - 1} \tag{1.151}$$

This is the result obtained by Einstein more then 10 years after Planck obtained his result, and 10 years before the advent of Quantum Physics and Second Quantization.

1.6 SELECTION RULES

Selection rules follow from the requirement that relevant matrix elements are non-zero in the application of Fermi's Golden Rule Equation (1.103). Besides doing explicit and occasionally time-consuming calculations of the matrix elements, it is sometimes possible to ascertain that they are zero by a conservation law. We will discuss conservation laws and their connection to symmetries in Chapter 3. But we are in a position to give some examples here using the results we obtained in this chapter.

The same selection rules apply to emission and absorption of photons because they are governed by the matrix element $\langle B|\mathcal{O}|A\rangle$ and its Hermitian conjugate where the operator $\mathcal{O}$ represents an electric or magnetic multipole moment.

An important selection rule is provided by the parity operator P. The parity operator P makes a reflection of a vector through the origin of a coordinate system so it changes $\mathbf{x} \to -\mathbf{x}$. It follows that under parity the momentum $\mathbf{p} \to -\mathbf{p}$, the angular momentum $\mathbf{L} = \mathbf{x} \times \mathbf{p} \to (-\mathbf{x}) \times (-\mathbf{p}) = \mathbf{L}$, etc. Two successive parity operations are equivalent to the unit operator so $P^2 = 1$. From the definition of the inverse of an operator we have $PP^{-1} = P^{-1}P = 1$ so we have $P = P^{-1}$. We require that the ket $|A_P\rangle = P|A\rangle$ that results from the parity operation has the same norm as the ket $|A\rangle$. Therefore we require $\langle A_P|A_P\rangle = \langle A|A\rangle$. But $\langle A_P| = \langle PA| = \langle A|\, P^\dagger$ so we have that $P^\dagger P = 1$. Because $P^{-1}P = 1$ it follows that $P^\dagger = P^{-1}$ or with $P^{-1} = P$ that the parity operator is Hermitian. Its eigenvalues are therefore real and observable and we write $P\,|A\rangle = \eta_P(A)\,|A\rangle$. Because $P^2 = 1$ we have that $\eta_P^2(A) = 1$ or $\eta_P(A) = \pm 1$. We talk about even and odd parity.

Now consider the transformation of the electric dipole moment under the parity operator $\mathbf{D} = \langle B|e\mathbf{x}|A\rangle \to \langle B_P|e(-\mathbf{x})|A_P\rangle = \eta_P^*(B)\,\eta_P(A)\langle B|e(-\mathbf{x})|A\rangle = -\eta_P(B)\eta_P(A)\,\langle B|e\mathbf{x}|A\rangle$. If the parity operator is a symmetry operator (physical properties of the system are the same before and after the parity operation) then we must require that $-\eta_P(B)\,\eta_P(A) = 1$ or $\eta_P(B) = -\eta_P(A)$ where we have used that η_P can only take on the values ± 1. Thus $E1$ transitions can only take place between states $|A\rangle$ and $|B\rangle$ if they have opposite parity. Conversely, if experiment shows that a certain transition is forbidden to go by $E1$, the two states in question must have the same parity or parity is not a good quantum number. We will see in Chapter 3 that parity is a good quantum number for electromagnetic and strong interactions (such as the nuclear force) but not for weak interactions (such as β-decay).

We next consider magnetic dipole transitions $M1$. As shown in the previous section they are governed by the magnetic dipole moment that transforms under the parity operator as $\boldsymbol{\mu} = \langle B|(e/2m)\mathbf{L}|A\rangle \to \langle B_p|(e/2m)(+\mathbf{L})|A_P\rangle = +\eta_P(B)\eta_P(A)\,\langle B|(e/2m)\mathbf{L}|A\rangle$. Note the absence of the $-$ sign that was present in the case of $E1$ transitions. Thus $M1$ transitions can only take place between states $|A\rangle$ and $|B\rangle$ if they have the same parity.

We next consider electric quadrupole transitions $E2$. As shown in the previous section they are governed by the electric quadrupole moment that transforms under the parity operator as $\langle B|e\,x_i x_j|A\rangle \rightarrow \langle B_p|e\,(-x_i)(-x_j)|A_P\rangle = +\eta_P(B)\eta_P(A)\langle B|e\,x_i x_j|A\rangle$. Thus $E2$ transitions can only take place between states $|A\rangle$ and $|B\rangle$ if they have the same parity.

A clear pattern is developing that we can generalize to higher multipole transitions.

We remind the reader that the parity of a state with orbital angular moment ℓ is equal to $\eta_P = (-1)^\ell$. We mentioned that the $2s \rightarrow 1s$ transitions are forbidden by $E1$ transitions using an argument based upon rotational symmetry. We now see that this transition is also forbidden to go by $E1$ by parity but allowed (by parity) to go with $M1$. However, the argument based upon rotational symmetry also forbids the transition by $M1$. Can you think of a way that the $2s \rightarrow 1s$ transition is allowed?

We now turn to selection rules having to do with angular momentum. In the considerations that follow we choose the z-axis, the axis of quantization, along $\mathbf{k}$, the momentum of the photon. We start again with electric dipole transitions $E1$. One can show that

$$[\mathbf{L}^2, [\mathbf{L}^2, x]] = 2\,(\mathbf{L}^2 x + x\,\mathbf{L}^2) \tag{1.152}$$

As we will see, this relation will lead to selection rules thanks to the fact that the operator whose matrix element we are interested in is one of the elements of the commutator, and to the fact that the commutator equals an expression that has that same operator on the right side. We call such commutation relations 'Magic'. It will not be necessary to point out the need to keep the order of $\mathbf{L}^2$ and x. The proof is a problem at the end of this chapter, which uses the fact that $\mathbf{x} \cdot \mathbf{L} = 0$. We can sandwich Equation (1.152) between the bra $\langle \ell', m'|$ and the ket $|\ell, m\rangle$

$$\langle \ell', m'|[\mathbf{L}^2, [\mathbf{L}^2, x]]|\ell, m\rangle = 2\,\langle \ell', m'|\mathbf{L}^2 x + x\,\mathbf{L}^2|\ell, m\rangle \tag{1.153}$$

The double commutator can be expanded as $\mathbf{L}^2\mathbf{L}^2 x - 2\,\mathbf{L}^2 x \mathbf{L}^2 + x\mathbf{L}^2\mathbf{L}^2$ and Equation (1.153) becomes

$$[\ell'(\ell'+1) - \ell(\ell+1)]^2\,\langle \ell', m'|x|\ell, m\rangle = 2\,[\ell'(\ell'+1) + \ell(\ell+1)]\,\langle \ell', m'|x|\ell, m\rangle \tag{1.154}$$

We conclude that for the dipole matrix element $\langle \ell', m'|ex|\ell, m\rangle$ to be non-zero we must have $[\ell'(\ell'+1) - \ell(\ell+1)]^2 - 2\,[\ell'(\ell'+1) + \ell(\ell+1)] = 0$. This equation must be solved for ℓ' for a given value of ℓ. We see right away that $\ell' = \ell = 0$ is a solution, but the electric dipole moment is zero for this combination, as pointed out before. To get the general solution we set $y = \ell'(\ell'+1)$ and $z = \ell(\ell+1)$ to get $(y-z)^2 - 2y - 2z = 0$ or $y = (z+1) \pm \sqrt{4z+1}$ or $\ell'(\ell'+1) = (\ell+1)(\ell+2)$ or $(\ell-1)\ell$. From this we get the two solutions $\ell' = \ell + 1$ and $\ell' = \ell - 1$. So the selection rule for $E1$ transitions is $\Delta\ell = \ell' - \ell = \pm 1$. This makes some but not complete sense from the point of view of conservation of angular momentum: we start with a state with orbital angular momentum ℓ and when we emit or absorb a photon with spin one we should reach final states with $\ell' = \ell - 1, \ell, \ell + 1$ according to the rules for combining two angular momenta. The combination with $\ell' = \ell$ is missing, as parity forbids it. We must take care to distinguish between orbital and intrinsic (spin) angular momentum. Such separation requires that one can go to the system's center-of-mass where the orbital angular momentum is zero leaving only the intrinsic angular momentum. This is not possible for a particle that travels with the speed of light and thus its orbital and intrinsic angular momenta cannot be separated.

We now consider L_z and remember that the z-axis, the axis of quantization, is parallel to **k**, the momentum of the photon. We consider again $E1$ transitions whose amplitude is proportional to $\langle B|e\mathbf{x}|A\rangle \cdot \boldsymbol{\varepsilon}$, see Equation (1.117) and Equation (1.118). Because $\boldsymbol{\varepsilon} \cdot \mathbf{k} = 0$ from Equation (1.19) we set $\boldsymbol{\varepsilon} = (\varepsilon_x, \varepsilon_y, 0)$ and we only have to consider $\langle B|ex|A\rangle$ and $\langle B|ey|A\rangle$. We introduce

$$x_{\pm} = x \pm iy \qquad \varepsilon_{\pm} = \varepsilon_x \pm i\varepsilon_y \tag{1.155}$$

Inverting these equations we find

$$\varepsilon_x = \frac{1}{2}(\varepsilon_+ + \varepsilon_-) \qquad x = \frac{1}{2}(x_+ + x_-) \tag{1.156}$$

$$\varepsilon_y = \frac{1}{2i}(\varepsilon_+ - \varepsilon_-) \qquad x = \frac{1}{2i}(x_+ - x_-) \tag{1.157}$$

Using the commutators

$$[L_z, x] = -\frac{y}{i} \qquad [L_z, y] = \frac{x}{i} \tag{1.158}$$

we can derive

$$[L_z, x_{\pm}] = \pm x_{\pm} \tag{1.159}$$

another 'magic relation'. After these preliminaries we return to $\langle B|ex|A\rangle$ and $\langle B|ey|A\rangle$ and consider a linear superposition of these two matrix elements. That involves $\varepsilon_x x + \varepsilon_y y = (\varepsilon_+ x_- + \varepsilon_- x_+)/2$ so we must study $\langle \ell', m'|\varepsilon_{\pm} x_{\mp}|\ell, m\rangle$. We do this as before by sandwiching the commutation relation in Equation (1.159) between the bra $\langle \ell', m'|$ and the ket $|\ell, m\rangle$

$$\langle \ell', m'|[L_z, x_{\pm}]|\ell, m\rangle = \pm\langle \ell', m'|x_{\pm}|\ell, m\rangle \tag{1.160}$$

From this relation we get

$$(m' - m)\langle \ell', m'|x_{\pm}|\ell, m\rangle = \pm\langle \ell', m'|x_{\pm}|\ell, m\rangle \tag{1.161}$$

or (reinserting $\varepsilon_{\mp}$)

$$(m' - m \mp 1)\langle \ell', m' | \varepsilon_{\mp} x_{\pm}] | \ell, m\rangle = 0 \tag{1.162}$$

Thus for the matrix element $\langle \ell', m' | \varepsilon_{\mp} x_{\pm} | \ell, m\rangle$ to be non-zero we require $m' - m \mp 1 = 0$ or $\Delta m = m' - m = \pm 1$. The polarization ε is associated with the absorption of a photon and ε^* is associated with the creation of a photon. We see from Equation (1.155) that $\varepsilon^*_{\pm} = \varepsilon_{\mp}$. Substitution of this relation in Equation (1.162) gives

$$(\Delta m \mp 1)\langle \ell', m' | \varepsilon^*_{\pm} x_{\pm} | \ell, m\rangle = 0 \tag{1.163}$$

For the upper signs in Equation (1.163) we find that for the matrix element $\langle \ell', m' | \varepsilon^*_{\pm} x_{\pm}] | \ell, m\rangle$ to be non-zero m' must be one unit larger than m. Conservation of the component of angular momentum along the axis of quantization **k** implies that the created photon has $S_z = -1$, that is, the created photon is a left-handed (LH) one. Thus $\varepsilon^*_{+} = \varepsilon_x - i\varepsilon_y$ is associated with the creation of a LH photon. Similarly we find using the lower sign in Equation (1.163) that $\varepsilon^*_{-} = \varepsilon_x + i\varepsilon_y$ is associated with the creation of a right-handed (RH) photon. Likewise, $\varepsilon_{+} = \varepsilon_x + i\varepsilon_y$ absorbs a RH photon while $\varepsilon_{-} = \varepsilon_x - i\varepsilon_y$ absorbs a LH photon. It is left to the reader to prove these statements, see Section 1.2.4 on Polarization and Spin.

We now turn to $M1$ transitions. From Equation (1.128) and Equation (1.129) we see that the relevant quantity is $(\mathbf{k} \times \boldsymbol{\varepsilon}) \cdot \langle \ell', m' | \mathbf{L} | \ell, m\rangle$. The component of **L** parallel to **k** does not contribute to the triple product, so we only need to consider the components of **L** that are perpendicular to **k**, the axis of quantization along which we choose the z-axis to be. Take for example $\mathbf{L} = (L_x, 0, 0) \propto (L_+ + L_-, 0, 0)$ with $L_{\pm}$ the raising and lowering operators. If we sandwich this between $\langle \ell', m'|$ and $|\ell, m\rangle$ we find that the matrix element $\langle \ell', m' | \mathbf{L} | \ell, m\rangle$ can be non-zero only if $m' = m \pm 1$. We can also sandwich the commutator relation $[\mathbf{L}^2, \mathbf{L}] = 0$ between that bra and ket to find that the matrix element $\langle \ell', m' | \mathbf{L} | \ell, m\rangle$ can be non-zero only if $\ell' = \ell$. We see from the examples given thus far that selection rules can be obtained by using magic commutation relations sandwiched between the bra $\langle \ell', m'|$ and the ket $|\ell, m\rangle$. Although a more systematic approach can be taken, we will not do that here because it is outside the scope of this book. Instead we will give a few more examples using the methods employed earlier in this section.

Consider for example $E2$ transitions. Among the nine quantities $\varepsilon_i k_j \langle B | x_i x_j | A\rangle$ to be considered, see Equation (1.133), we chose $\langle B | xy | A\rangle$. One can easily show that $xy = (x^2_+ - x^2_-)/(4i)$ and that

$$[L_z, x^2_{\pm}] = \pm 2x^2_{+} \tag{1.164}$$

(another magic relation) where $x_\pm$ are defined in Equation (1.155). Sandwiching the commutator relation in Equation (1.164) gives

$$(m' - m \mp 2)\langle \ell', m' | x_\pm^2 | \ell, m\rangle = 0 \tag{1.165}$$

which leads to the selection rule $\Delta m = m' - m = \pm 2$. Similarly, the easily derived commutation relation

$$[L_z, x_\pm z] = \pm x_\pm z \tag{1.166}$$

(another magic relation) leads to

$$(m' - m \mp 1)\langle \ell', m' | x_\pm z | \ell, m\rangle = 0 \tag{1.167}$$

and the selection rule is $\Delta m = \pm 1$. The latter type of matrix elements appear when the photon is emitted along the z-axis so only terms with ε_1 and ε_2 and thus with xz and yz in $\varepsilon_i k_j \langle B|x_i x_j|A\rangle$ contribute. We see that only $\Delta m = \pm 1$ are possible and $\Delta m = \pm 2$ are forbidden in this special case. This is of course due to conservation of angular momentum and the photon having $S_z = \pm 1$ along $\mathbf{k}$, the axis of quantization. Only when the photon's momentum is not along the z-axis do we get terms like xy and $\Delta m = \pm 2$. This all hangs together very well!

The selection rules for ℓ in $E2$ can be obtained by using the fact that products $x_i x_j$ can be written in terms of spherical harmonics and calling upon the addition theorem of spherical harmonics. For example $x \propto rY_1^0$ and $y \propto rY_1^1$. We find that $\Delta\ell = \pm 2$ transitions are allowed, as are $\Delta\ell = \pm 1$. Thus the selection rules $\Delta\ell = \pm 1$ and $\Delta m = 0, \pm 1$ that you have been asked to memorize without derivation are not strictly true. They only apply to $E1$ transitions, the dominant ones among multipole transitions, see Section 1.4.

More generally, when considering matrix elements of the type $\langle B|\mathcal{O}|A\rangle = \langle \ell_B, m_B|\mathcal{O}|\ell_A, m_A\rangle$ we know that the ket $|A\rangle = |\ell_A, m_A\rangle$ transforms as an object with angular momentum ℓ_A, while $\mathcal{O} = \mathbf{x}$ or $\mathbf{L}$ transforms as an $\ell_\mathcal{O} = 1$ object. Likewise the $\mathcal{O} = x_i x_j$ transforms as an object with $\ell_\mathcal{O} = 2$. The rule for combining two objects with angular momenta ℓ_A and $\ell_\mathcal{O}$ specifies that the resulting total angular momentum can take on a value from the list $|\ell_\mathcal{O} - \ell_A|$, $|\ell_\mathcal{O} - \ell_A| + 1$, $|\ell_\mathcal{O} - \ell_A| + 2$, $\ldots$ $|\ell_\mathcal{O} + \ell_A|$. From among these the bra $\langle B| = \langle \ell_B, m_B|$ projects out the value ℓ_B if it is present in the list. This leads to the selection rules $\Delta\ell = 0, \pm 1$ for $E1$ and $M1$, $\Delta\ell = 0, \pm 1, \pm 2$ for $E2$ and $M2$, etc. Of course parity conservation eliminates some of these possibilities, as discussed at the beginning of this section.

PROBLEMS

(1) *Gauge Transformations.*
Consider, in a finite volume of space, a superposition of a Coulomb field from a stationary point charge and a time-dependent electromagnetic field. There are no currents, the time-dependent fields are generated by currents at infinity. Use the Lorenz gauge. The potentials are given by:

$$\mathbf{A}(\mathbf{x}, t) = \mathbf{a} \sin(\mathbf{k} \cdot \mathbf{x} - \omega t) \tag{1.168}$$

$$\phi(\mathbf{x}, t) = v \sin(\mathbf{k} \cdot \mathbf{x} - \omega t) + \frac{e}{|\mathbf{x}|} \tag{1.169}$$

(a) Derive relations between $\mathbf{a}$, v, $\mathbf{k}$ and ω, using the wave equations and the Lorenz condition and comment on these.
Make a gauge transformation to the radiation gauge, using

$$\chi(\mathbf{x}, t) = \frac{v}{\omega} \cos(\mathbf{k} \cdot \mathbf{x} - \omega t) \tag{1.170}$$

(b) Calculate the new potentials $\mathbf{A}$ and ϕ.
(c) Show that the new $\mathbf{A}$ is (still) perpendicular to $\mathbf{k}$.
(d) Write down the Lorenz condition in terms of the new potentials.
(e) Briefly comment on your results: does the presence of a static Coulomb field change any results obtained thus far in second quantization of the electromagnetic field?

(2) *Second Quantization.*
(a) Derive the momentum operator of the quantized electromagnetic field, starting with the Poynting vector. Show all steps.
(b) Consider this result and the energy operator. What were Planck's and Einstein's hypotheses concerning blackbody radiation and the photoelectric effect and how can these be 'derived' from the results of second quantization?

(3) *Spontaneous Emission.*
Calculate the rate for spontaneous emission of a single photon by a one-dimensional harmonic oscillator making a transition from state $|i\rangle$ to state $|f\rangle$. Keep it general: do not assume that one of the states is the ground state or that they correspond to adjacent energy levels. Sum over the polarization of the emitted photon to obtain the rate differential in the angles of the two photons. Integrate the differential probability over angles to get the total rate. Use the operator formalism instead of wavefunctions.

Do the calculation seperately in:

(a) The electric dipole approximation.

(b) The electric quadrupole approximation.

(c) Summarize the selection rules governing these two transitions, including parity.

(4) *Two-Photon Emission.*

Consider $2s \rightarrow 1s$ transitions in the hydrogen atom. Ignore the spin of the electron and proton.

(a) What is the rate for single photon emission from the $\mathbf{A} \cdot \mathbf{p}$ perturbative term in the Hamiltonian in the electric dipole approximation?

(b) Would it help to consider magnetic dipole or electric quadrupole or even higher multipolarities?

There is a $\mathbf{A}^2$ perturbative term in the Hamiltonian.

(c) Show qualitatively that this term can change the net number of photons between the initial and final state by $-2, 0, +2$.

(d) Derive the Golden Rule for the case of spontaneous two-photon emission. Remember to include phase space factors for two photons.

(e) Discuss the salient features of your result. Is energy conserved? What is the power of the fine structure constant in your answer? Compare with the result for single photon emission.

(f) Use the electric dipole approximation and sum over the polarizations of the two photons to obtain the rate differential in the angles of emission of the two photons.

(g) Integrate over the angles to get the total rate. Get a numerical estimate and compare with the result for single photon emission in the electric dipole approximation.

(5) *Polarization.*

Consider the two-photon state

$$a_\lambda^\dagger(\mathbf{k})a_\mu^\dagger \mathbf{l}|0\rangle \tag{1.171}$$

(a) Use the expressions given in class for the number operator N, the Hamiltonian H, and the momentum operator $\mathbf{P}$ to calculate the eigenvalues of these three operators. Show all steps, no matter how trivial.

Consider the following two expressions:

$$|1\rangle = a_1^\dagger(\mathbf{k})e^{-i(\mathbf{k}\cdot\mathbf{x}-\omega t)}|0\rangle$$

$$|2\rangle = a_2^\dagger(\mathbf{k})e^{-i(\mathbf{k}\cdot\mathbf{x}-\omega t)}|0\rangle$$

(b) What is the physical meaning of these two states? Be sure to specify the polarization of each of the two states.
(c) Specify the polarization of the state $|1\rangle + |2\rangle$.
(d) Define a state that corresponds to a single photon that is linearly polarised along an axis that makes a 30 degree angle with the x-axis and a 60 degree angle with the negative y-axis.
(e) Specify the polarization of the state $|1\rangle + i|2\rangle$.
(f) Same question for the state $|1\rangle - i|2\rangle$.
(g) Define a creation operator $a_R^\dagger(\mathbf{k})$ for creating a RH polarized photon.

(6) *Stark Effect.*
A hydrogen atom in its ground state is placed in a spatially homogeneous time-dependent electric field $\mathcal{E} = 0$ for $t < 0$ and $\mathcal{E} = \mathcal{E}_0 \exp(-t/\tau)$ for $t > 0$. Find the probability, in lowest order of perturbation theory, that after a sufficiently long time the atom is in the $2p$ state in which the component of the orbital angular momentum in the direction of $\mathcal{E}$ is zero.

(7) *Commutation Relations and Selection Rules.*
Derive the relation:

$$[\mathbf{L}^2, [\mathbf{L}^2, \mathbf{x}]] = 2(\mathbf{L}^2\mathbf{x} + \mathbf{x}\mathbf{L}^2) \tag{1.172}$$

This is a relation involving the angular momentum operator $\mathbf{L}$ and the position vector $\mathbf{x}$. For simplicity it should be proven seperately for each of the three components of $\mathbf{x}$. The following steps will be helpful. If you have trouble, prove things one component at a time and then generalize.

(a) Show that if in an operator product P and Q are exchanged the following relation holds:

$$AB\ldots PQ\cdots Z = AB\cdots QP\cdots Z + AB\cdots[P,Q]\cdots Z \tag{1.173}$$

(b) Show that

$$[A, BC] = B[A, C] + [A, B]C \tag{1.174}$$

where A, B and C are operators.

(c) Show that

$$[A^2, B] = A[A, B] + [A, B]A \tag{1.175}$$

where A, B and C are operators.

(d) Show that

$$[L_i, x_j] = i\varepsilon_{ijk}x_k \tag{1.176}$$

where x_i is the i-th component of the vector $\mathbf{x}$ and ε_{ijk} the totally anti-symmetric (Levi-Civita) tensor.

(e) Show that

$$[\mathbf{L}^2, x_i] = 2i\varepsilon_{ijk}x_jL_k + 2x_i \qquad (1.177)$$

The next item is for extra credit. I advise you to attempt it after you have finished the rest of the assignment and only if you can spare the time.

(f) Show that

$$\left[\mathbf{L}^2, [\mathbf{L}^2, x]\right] = 2(\mathbf{L}^2 x + x\mathbf{L}^2) \qquad (1.178)$$

Generalize this to obtain the relation at the beginning of this problem.

(8) *M1 and E2 Radiation.*

(a) Complete the calculation for spontaneous emission by M1 in the manner it was done for E1.

(b) Same as (a) but for E2.

(9) *Spontaneous and Induced Transitions.*
Consider transitions between states of a rotator whose Hamiltonian is

$$H = \frac{\vec{L}^2}{2I}$$

The rotor has an electric dipole moment $\vec{D}$ directed along its axis of rotation and an electric quadrupole moment given by $Q_{ij} = ex_ix_j$.

(a) Solve for the eigenvalues and eigenkets of H.

(b) What is the rate for spontaneous single photon emission in the electric dipole approximation?

(c) Derive the angular momentum and parity selection rules that govern the $E1$ transitions of b).

(d) Same questions as in (b) and (c) but now for electric quadrupole radiation $E2$.

(e) Write down the Hamiltonian of the system in the presence of a spatially homogeneous time-dependent electric field. Consider both the electric dipole and electric quadrupole moments. Assume that the time dependence is slow enough that induced magnetic fields may be ignored.

(f) With the system in its ground state, a spatially homogeneous time-dependent electric field is turned on at $t = 0$. The electric field is given by $\mathcal{E} = \mathcal{E}_0 \sin \omega t$ for $0 < t < 2\pi/\omega$ and $\mathcal{E} = 0$ at all other values of t. Find the probability, in lowest order of perturbation theory, that after a sufficiently long (but not infinite) time the rotator is in a given excited state.

(g) What selection rules, if any, apply in the transitions in (f)?

2

Scattering

2.1 SCATTERING AMPLITUDE AND CROSS SECTION

Scattering of one object by another plays an important role in the sciences and beyond. The scattering of a photon by atoms or molecules makes it possible for you to read this text. Photons from ambient light sources are scattered from the molecules on the surface of the page, some of these enter the eye and are again scattered by molecules in the retina. In physics, scattering of particles (often photons) off molecules, atoms, nuclei, or elementary particles give information about the target's constituents. We shall see that a necessary condition for 'looking inside' the target (we call that 'probing the target') requires that the projectile particles have a wavelength much shorter than the size of the object to be probed. Quantum Physics tells us that $\lambda = h/p$ so to get short wavelengths we need large momenta and thus energies. This is one of the *two* reasons for having ever larger accelerators in elementary particle physics where we are now probing the constituents ('quarks') of the proton and the neutron. Of course this works best if the probe itself is a structureless particle (or has a 'known' structure). The second reason is that scattering experiments give information on the interaction between projectile and target. If a new type of interaction is at work in a scattering process, measurements of the scattering process will give information about that interaction. The scattering process is shown schematically in Figure 2.1. A beam of incident particles (the projectiles), each with momentum **k** parallel to the $+z$-axis, is incident from the left upon a target at rest. The target is surrounded by one or more detectors that detect scattered particles within a usually small solid angle subtended by each detector. The incident particle is represented by a plane wave $\psi_i = \exp(ikz)$ while the scattered particle is represented by a spherical wave $\psi_s = [f(\theta, \phi)\exp(ikr)]/r$ with **k** the wave vector. The angles θ and ϕ specify

An Introduction to Advanced Quantum Physics Hans P. Paar

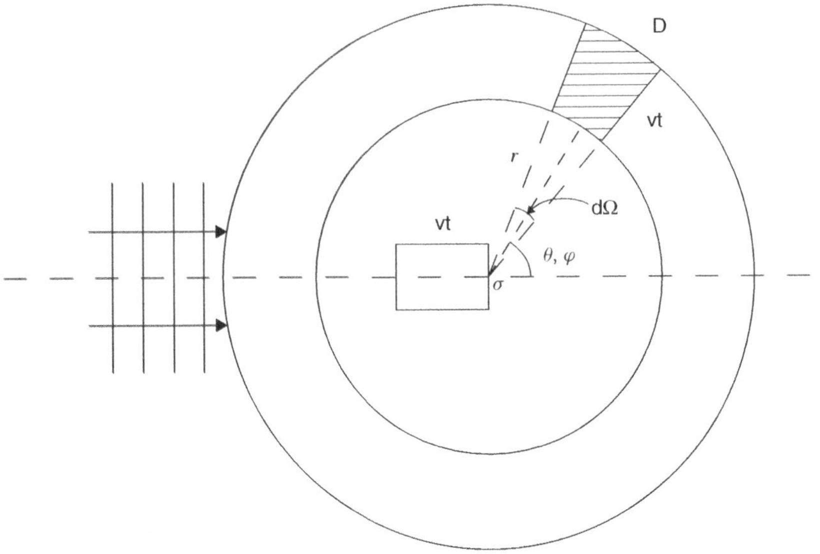

Figure 2.1 Schematic representation of a scattering process

the angular position of the detector, r is the distance between the target and the detector. A time dependence of the form $\exp(-i\omega t)$ is understood, so the incident wave is moving in the $+\mathbf{k}$ direction while the scattered spherical wave is moving in the $+r$ direction. The $1/r$ dependence of the scattered wave is chosen so that $|\psi_s|^2$ drops off with r as $1/r^2$. Thus $|\psi_s|^2$ integrated over the surface of a sphere with radius r is independent of the value of r as it should be. This can also be seen using the probability current in spherical coordinates.

The complete wavefunction is

$$\psi(\mathbf{x}) = \psi_i + \psi_s = e^{i\mathbf{k}\cdot\mathbf{x}} + f(\theta,\phi)\frac{e^{ikr}}{r} \tag{2.1}$$

The quantity $f(\theta,\phi)$ is called the scattering amplitude. Assume that the scattering process takes place in a large box with volume V. We want to find an expression for the probability for an interaction between the projectile and the target. Later we will set that probability equal to w_{fi}, the probability for the system to go from the initial state before the scattering $|i\rangle$ to the final state after the scattering $|f\rangle$. Assume that the interaction effectively gives the target an area σ transverse to the incident particle's direction. We call σ the 'cross-section' of the target. If the incident particle has a velocity v (parallel to the z-axis) and we observe the system for an infinitely short time t then for the projectile to reach the target in that time interval, the incident particle must be in a right cylinder with base σ and length vt with the target at the right end of the cylinder. This is shown in Figure 2.1. The probability P to find the incident particle in the cylinder is

$$P = \frac{v\,t\,\sigma\,|\psi_i|^2}{V} = \frac{v\,t\,\sigma}{V} \tag{2.2}$$

The probability P that the scattered particle passes through the surface of a sphere at radius r is

$$P = \frac{v\,t\,r^2 \int d\Omega\,|\psi_s|^2}{V} = \frac{v\,t}{V} \int d\Omega\,|f(\theta,\phi)|^2) \tag{2.3}$$

In the steady state where a steady stream of incident particles is scattered, the two probabilities in Equation (2.2) and Equation (2.3) are equal because of conservation of probability (incident particles are neither created nor annihilated in the scattering) and we get

$$\sigma = \int d\Omega\,|f(\theta,\phi)|^2 \tag{2.4}$$

But $\sigma = \int (d\sigma/d\Omega)\,d\Omega$ so

$$\frac{d\sigma}{d\Omega} = |f(\theta,\phi)|^2 \tag{2.5}$$

The scattering amplitude $f(\theta,\phi)$ and the Hamiltonian are connected in that we can calculate the scattering amplitude if we know the Hamiltonian, or if we do not know the Hamiltonian, knowledge of the scattering amplitude can provide guidance toward formulating the Hamiltonian. We can also connect the probability of Equation (2.2) with the transition probability per unit time w_{fi} for the projectile-target system to go from the initial state $|i\rangle$ (before the scattering) to the final state $|f\rangle$ (after the scattering)

$$w_{fi} = \frac{dP}{dt} \tag{2.6}$$

Using Equation (2.2) in Equation (2.6) we get

$$\sigma = \frac{w_{fi}\,V}{v} \tag{2.7}$$

We shall see that V cancels in actual calculations and that often $v = c = 1$ when the target is stationary. We see from Equation (2.7) that cross-section and transition probability are the same apart from trivial factors.

2.2 BORN APPROXIMATION

2.2.1 Schrödinger Equation

The Born approximation is a method that allows us to solve the Schrödinger equation for the wavefunction ψ_s of the scattered particle. We must find solutions of the time-independent Schrödinger equation (because the scattering is steady state)

$$\nabla^2\psi + k^2\psi = 2mV(\mathbf{x})\,\psi \tag{2.8}$$

with $k^2 = 2mE = \mathbf{k}^2$ and $V(\mathbf{x})$ the potential energy of the projectile in the presence of the target. We will assume that the scattering potential is weak so that the scattered wave has a small amplitude relative to the amplitude of the incoming wave. We try a solution of the form Equation (2.1)

$$\psi(\mathbf{x}) = e^{i\mathbf{k}\cdot\mathbf{x}} + \psi_s \tag{2.9}$$

Substitution of Equation (2.9) in Equation (2.8) and using the fact that

$$(\nabla^2 + k^2)\,e^{i\mathbf{k}\cdot\mathbf{x}} = 0 \tag{2.10}$$

gives

$$(\nabla^2 + k^2)\,\psi_s = 2mV(\mathbf{x})\,\psi \tag{2.11}$$

Note that the left-hand side of Equation (2.11) has ψ_s in it as we want, but that the right-hand side has ψ in it and thus depends upon ψ_s. Ignoring that complication for the moment, we solve the linear inhomogeneous differential Equation (2.11) in the usual fashion by adding the general solution of the homogeneous equation to a particular solution of the inhomogeneous equation. One of the two solutions of the homogeneous equation is $\exp(i\mathbf{k}\cdot\mathbf{x})$, see Equation (2.10); the other one is $\exp(-i\mathbf{k}\cdot\mathbf{x})$ and is discarded because it corresponds to incident particles moving in the wrong direction. We will find a particular solution next.

2.2.2 Green's Function Formalism

To find a particular solution of Equation (2.11) we make use of the Green's function formalism. You will be familiar with this formalism from the study of Poisson's equation in electrostatics where solutions of the Poisson equation were found using the Green's function formalism. You may want to review this in preparation for the present application. We can put Equation (2.11) in the standard form for the use of the Green's function formalism by setting

$$2mV(\mathbf{x})\,\psi = -4\pi\rho(\mathbf{x}) \tag{2.12}$$

in Equation (2.11). It then becomes

$$(\nabla^2 + k^2)\,\psi_s = -4\pi\rho(\mathbf{x}) \tag{2.13}$$

To find solutions to this equation we first study

$$(\nabla^2 + k^2)\,G(\mathbf{x},\mathbf{x}') = -4\pi\,\delta^3(\mathbf{x}-\mathbf{x}') \tag{2.14}$$

where $G(\mathbf{x},\mathbf{x}')$ is called the Green's function. We encounter this equation in electrostatics without the k^2 term. It is then called Poisson's equation. If we know the solution $G(\mathbf{x},\mathbf{x}')$ of Equation (2.14) we obtain the solution of Equation (2.13) as

$$\psi_s(\mathbf{x}) = \int d^3\mathbf{x}'\,G(\mathbf{x},\mathbf{x}')\,\rho(\mathbf{x}') \tag{2.15}$$

It is seen that we subdivide the function $\rho(\mathbf{x})$ in Equation (2.13) into spatially infinitesimal small pieces (represented by the δ^3-function in Equation (2.14)), solve the equation for one such small piece, and afterward sum (integrate) the piecemeal solutions for all small pieces of $\rho(\mathbf{x})$. For the formal proof we substitute Equation (2.15) in Equation (2.13) and use Equation (2.14)

$$\begin{aligned}(\nabla^2 + k^2)\,\psi_s &= (\nabla^2 + k^2)\int d^3\mathbf{x}'\,G(\mathbf{x},\mathbf{x}')\,\rho(\mathbf{x}')\\ &= \int d^3\mathbf{x}'\,(\nabla^2 + k^2)\,G(\mathbf{x},\mathbf{x}')\,\rho(\mathbf{x}')\\ &= \int d^3\mathbf{x}'[-4\pi\,\delta^3(\mathbf{x}-\mathbf{x}')]\,\rho(\mathbf{x}')\\ &= -4\pi\,\rho(\mathbf{x})\end{aligned} \tag{2.16}$$

So far the procedure follows the one used in electrostatics. It was shown there that the Green's function $G(\mathbf{x},\mathbf{x}') = 1/|\mathbf{x}-\mathbf{x}'|$ is a solution of Equation (2.14) if $k^2 = 0$. In the present case $k^2 \neq 0$ and the Green's function that is a solution of Equation (2.14) is

$$G(\mathbf{x},\mathbf{x}') = \frac{e^{ik|\mathbf{x}-\mathbf{x}'|}}{|\mathbf{x}-\mathbf{x}'|} \tag{2.17}$$

To prove this we first note that $(\nabla^2 + k^2)\exp(ikr)/r = 0$ for all values of k if $r = |\mathbf{x}-\mathbf{x}'| \neq 0$. This can be seen by writing the ∇^2 operator in spherical coordinates as

$$\nabla^2 = \frac{1}{r^2}\frac{\partial}{\partial r}\left(r^2\frac{\partial}{\partial r}\right) + \left(\frac{\partial}{\partial\theta},\frac{\partial}{\partial\phi}\right) - \text{ terms} \tag{2.18}$$

Using this we find

$$\nabla^2 \frac{e^{ikr}}{r} = \frac{1}{r^2} \frac{\partial}{\partial r} \left[r^2 \frac{\partial}{\partial r} \frac{e^{ikr}}{r} \right] = -\frac{k^2}{r} e^{ikr} \tag{2.19}$$

if $r \neq 0$. The terms with the derivatives in θ and ϕ in Equation (2.18) do not contribute because there is no angular dependence in Equation (2.17). The result in Equation (2.19) precisely cancels the second term on the left-hand side of Equation (2.14). Because $G(\mathbf{x}, \mathbf{x}')$ is singular for $r = 0$ we must consider this case separately, analogous to the case for the Poisson equation in electrostatics. We need to show that $(\nabla^2 + k^2)\, G(\mathbf{x}, \mathbf{x}')$ behaves like $-4\pi\, \delta^3(\mathbf{x} - \mathbf{x}')$ when $\mathbf{x} - \mathbf{x}' \approx 0$. In particular, integrating $(\nabla^2 + k^2)\, G(\mathbf{x}, \mathbf{x}')$ over an infinitely small volume around $r = |\mathbf{x} - \mathbf{x}'| \approx 0$ should give -4π. The k^2 term does not contribute, as can be seen by expressing the integral in spherical coordinates

$$\int k^2 \frac{e^{ikr}}{r} r^2 dr \sin\theta\, d\theta\, d\phi = 0 \text{ as } r \to 0 \tag{2.20}$$

To evaluate the ∇^2 term we use the $\mathbf{\nabla}$ operator in spherical coordinates

$$\mathbf{\nabla}\left(\frac{e^{ikr}}{r}\right) = \mathbf{e}_r \frac{\partial}{\partial r}\left(\frac{e^{ikr}}{r}\right) + (\mathbf{e}_\theta, \mathbf{e}_\phi)\text{–terms} = \mathbf{e}_r \left(\frac{ikr - 1}{r^2}\right) e^{ikr} \tag{2.21}$$

because the $\mathbf{e}_\theta$ and $\mathbf{e}_\phi$ terms do not contribute (there is no angular dependence). Here $\mathbf{e}_r$, $\mathbf{e}_\theta$ and $\mathbf{e}_\phi$ are mutually perpendicular unit vectors in the r, θ and ϕ direction. We substitute Equation (2.21) in Gauss' Law

$$\int d^3\mathbf{x}\, \mathbf{\nabla} \cdot (\mathbf{\nabla} f) = \oint \mathbf{n} \cdot (\mathbf{\nabla} f)\, dS \tag{2.22}$$

to get, letting $r \to 0$ when appropriate

$$\begin{aligned}
\int d^3\mathbf{x}\, \nabla^2 \left(\frac{e^{ikr}}{r}\right) &= \int d^3\mathbf{x}\, \mathbf{\nabla} \cdot \mathbf{\nabla}\left(\frac{e^{ikr}}{r}\right) \\
&= \oint \frac{\mathbf{r}}{r} \cdot \mathbf{e}_r \left(\frac{ikr - 1}{r^2}\right) e^{ikr}\, dS \\
&= \int \left(\frac{-1}{r^2}\right) r^2\, d\Omega \\
&= \int (-1)\, d\Omega \\
&= -4\pi
\end{aligned} \tag{2.23}$$

where in the second line we have used that $\mathbf{n} = \mathbf{r}/r$ and in the third line we have neglected ikr relative to 1 and set $\exp(ikr) = 1$. Thus we have shown that $(\nabla^2 + k^2)\exp(ikr)/r$ behaves like $-4\pi\,\delta(r)$ with $r = |\mathbf{x} - \mathbf{x}'|$ or that Equation (2.17) is a solution of Equation (2.14) for all r. In electrostatics with $k = 0$ the Green's function in Equation (2.17) becomes $1/|\mathbf{x} - \mathbf{x}'|$ as it should be for the potential of an electric charge. So if you did not study the Green's function formalism in electrostatics, you now know enough to solve Poisson's equation as well.

2.2.3 Solution of the Schrödinger Equation

We can now write down the solution of Equation (2.11) using Equation (2.15) and Equation (2.17) and the definition in Equation (2.12) of $\rho(\mathbf{x})$

$$\psi_s(\mathbf{x}) = -\frac{1}{4\pi}\,2m\int d^3\mathbf{x}'\,\frac{e^{ik|\mathbf{x}-\mathbf{x}'|}}{|\mathbf{x}-\mathbf{x}'|}\,V(\mathbf{x}')\,\psi(\mathbf{x}') \tag{2.24}$$

and this expression for $\psi_s(\mathbf{x})$ can be substituted in Equation (2.9) to get

$$\psi(\mathbf{x}) = e^{i\mathbf{k}\cdot\mathbf{x}} - \left(\frac{2m}{4\pi}\right)\int d^3\mathbf{x}'\,\frac{e^{ik|\mathbf{x}-\mathbf{x}'|}}{|\mathbf{x}-\mathbf{x}'|}\,V(\mathbf{x}')\,\psi(\mathbf{x}') \tag{2.25}$$

The solution is not really a solution because the unknown function $\psi(\mathbf{x})$ appears on both sides of Equation (2.25). Note the manner in which $\mathbf{x}$ and $\mathbf{x}'$ appear in Equation (2.25). We see that if $V(\mathbf{x}) = 0$ then $\psi_s(\mathbf{x}) = 0$ and thus $f(\theta, \phi) = 0$ in Equation (2.1) and $\psi(\mathbf{x}) = \exp(i\mathbf{k}\cdot\mathbf{x})$, representing the incident plane wave without any scattering. This gives us a clue on how to proceed iteratively when $V(\mathbf{x})$ is not zero but 'small'.

A small $V(\mathbf{x})$ leads to a small $\psi_s(\mathbf{x})$ relative to $\exp(i\mathbf{k}\cdot\mathbf{x})$. The first iteration consists of substituting $\psi(\mathbf{x})$ of Equation (2.25) in place of ψ in the integrand on the right-hand side of Equation (2.25) and calculating a new and improved $\psi(\mathbf{x})$. The second iteration is obtained when we substitute this new $\psi(\mathbf{x})$ in place of ψ in the integrand and calculate a newer further improved $\psi(\mathbf{x})$ and substituting it in the integrand, and so on. Each step in this iterative process should improve $\psi(\mathbf{x})$ over the result of the previous step as long as $V(\mathbf{x})$ is small and the procedure converges. This might be expected to be the case because each step increases the power of $V(\mathbf{x})$ by one and we assumed that $V(\mathbf{x})$ is 'small'. During this process we introduce an additional integration for each iteration. Each integral must have its own integration variable. To this end we rename the integration variable $\mathbf{x}'$ in Equation (2.25) $\mathbf{x}_1$ to get

$$\psi(\mathbf{x}) = e^{i\mathbf{k}\cdot\mathbf{x}} + \left(-\frac{2m}{4\pi}\right)\int d^3\mathbf{x}_1\,G(\mathbf{x},\mathbf{x}_1)\,V(\mathbf{x}_1)\,\psi(\mathbf{x}_1) \tag{2.26}$$

where we have introduced the shorthand $G(\mathbf{x}, \mathbf{x}')$ for Green's function. We replace the integration variable $\mathbf{x}_1$ by $\mathbf{x}_2$ and replace $\mathbf{x}$ by $\mathbf{x}_1$ in Equation (2.26) to obtain

$$\psi(\mathbf{x}_1) = e^{i\mathbf{k}\cdot\mathbf{x}_1} + \left(-\frac{2m}{4\pi}\right) \int d^3\mathbf{x}_2\, G(\mathbf{x}_1, \mathbf{x}_2)\, V(\mathbf{x}_2)\, \psi(\mathbf{x}_2) \tag{2.27}$$

We now substitute the expression in Equation (2.27) for $\psi(\mathbf{x}_1)$ in the integrand on the right-hand side of Equation (2.26) and get

$$\begin{aligned}\psi(\mathbf{x}) = e^{i\mathbf{k}\cdot\mathbf{x}} &+ \left(-\frac{2m}{4\pi}\right) \int d^3\mathbf{x}_1\, G(\mathbf{x}, \mathbf{x}_1)\, V(\mathbf{x}_1)\, e^{i\mathbf{k}\cdot\mathbf{x}_1} \\ &+ \left(-\frac{2m}{4\pi}\right)^2 \int d^3\mathbf{x}_1\, d^3\mathbf{x}_2\, G(\mathbf{x}, \mathbf{x}_1)\, V(\mathbf{x}_1)\, G(\mathbf{x}_1, \mathbf{x}_2)\, V(\mathbf{x}_2)\, \psi(\mathbf{x}_2)\end{aligned} \tag{2.28}$$

When repeating this procedure one obtains an infinite series called the von Neumann series

$$\begin{aligned}\psi(\mathbf{x}) = e^{i\mathbf{k}\cdot\mathbf{x}} &+ \left(-\frac{2m}{4\pi}\right) \int G(\mathbf{x}, \mathbf{x}_1)\, [V(\mathbf{x}_1)\, d^3\mathbf{x}_1]\, e^{i\mathbf{k}\cdot\mathbf{x}_1} \\ &+ \left(-\frac{2m}{4\pi}\right)^2 \int G(\mathbf{x}, \mathbf{x}_1)[V(\mathbf{x}_1)\, d^3\mathbf{x}_1] G(\mathbf{x}_1, \mathbf{x}_2)[V(\mathbf{x}_2)\, d^3\mathbf{x}_2]\, e^{i\mathbf{k}\cdot\mathbf{x}_2} + \cdots\cdots\end{aligned} \tag{2.29}$$

If one truncates the series after the second term we speak of the Born approximation. There is an elegant graphical way to represent the terms in this series. The method uses three building blocks shown in Figure 2.2. The first building block shown on top is the graphical element for $\exp(i\mathbf{k} \cdot \mathbf{x})$, a line that ends at the point $\mathbf{x}$. It represents an incident particle propagating to the point $\mathbf{x}$ without any scattering. The building block shown next is the graphical element for $G(\mathbf{x}_1, \mathbf{x}_2)$, a line segment between the points $\mathbf{x}_2$ and $\mathbf{x}_1$. It represents the propagation of a particle from $\mathbf{x}_2$ to $\mathbf{x}_1$. The last building block shown is the graphical element for $V(\mathbf{x})\, d^3\mathbf{x}$, a bubble. It represents a single scatter at $\mathbf{x}$. The factors in the terms of the von Neumann series Equation (2.29) are taken from right to left (as we are used to in Quantum Physics). We show in Figure 2.3 the graphs for the terms in the Neumann series. The first graph shown on top of Figure 2.3 represents the first term in Equation (2.29) for the incident particle that propagates without scattering to the point $\mathbf{x}$. The second graph represents the second term in Equation (2.29) where the incident particle is propagated to the point $\mathbf{x}_1$, scattered at $\mathbf{x}_1$ by $V(\mathbf{x}_1)\, d^3\mathbf{x}_1$, and propagated from $\mathbf{x}_1$ to $\mathbf{x}$. The net result is that the incident particle is propagated to the space point $\mathbf{x}$ with a single scatter in the target along the way. The third graph in Figure 2.3 represents the third term in Equation (2.29) where the incident particle is propagated to the space point $\mathbf{x}_2$, scattered at $\mathbf{x}_2$ by $V(\mathbf{x}_2)\, d^3\mathbf{x}_2$, propagated

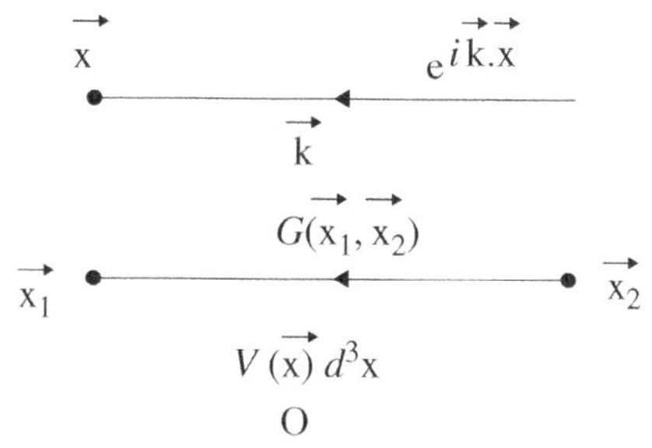

Figure 2.2 Building blocks for the graphical representation of terms in the von Neuman series

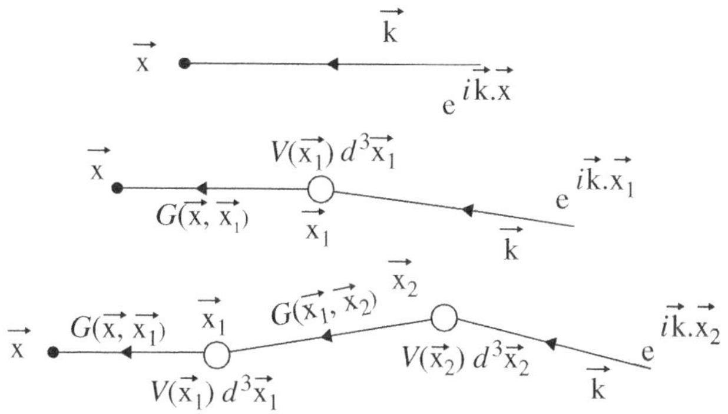

Figure 2.3 Graphical representation of the first three terms in the von Neumann series

from $\mathbf{x}_2$ to $\mathbf{x}_1$, scattered at $\mathbf{x}_1$ by $V(\mathbf{x}_1)\, d^3\mathbf{x}_1$, and propagated from $\mathbf{x}_1$ to $\mathbf{x}$. The net result is that the incident particle is propagated to the point $\mathbf{x}$, but now with two scatters in the target along the way. The figure represents all of space but in practice the points $\mathbf{x}_i$ are all near the target because only there is $V(\mathbf{x}_i)$ non-zero. As noted before, under our assumption that V is sufficiently small, such double scattering is less likely than single scattering. The manner in which these graphical representations can be extended to the next terms is clear. Note that the wavefunction $\psi(\mathbf{x})$ is the sum of terms representing no scatter, one scatter, two scatters, and so on. We sum wave function components (or amplitudes) over all possible paths through the target. Note that we do *not* sum probabilities.

Our current treatment is non-relativistic. This type of graphical representation has been refined by Feynman for the relativistically correct theory of the interaction of the electromagnetic field with charged particles. That theory is called Quantum Electrodynamics (QED) and its graphs representing the interaction of a charge with the electromagnetic field are called Feynman graphs.

2.2.4 Born Approximation

We often limit ourselves to the Born approximation and use the second term in Equation (2.29) for ψ_s. It represents the scattered particle and we use it to calculate the amplitude for finding the scattered particle in a detector located at the position $\mathbf{x}$. Detectors are placed at distances *much* larger than the typical dimensions of the target. We are therefore interested in the limit that $|\mathbf{x}| \to \infty$ while $\mathbf{x}'$ remains finite and *very* small relative to $|\mathbf{x}|$ so $|\mathbf{x}'|/|\mathbf{x}| \ll 1$. We expand the quantity $|\mathbf{x}-\mathbf{x}'|$ in Equation (2.17) as follows

$$\begin{aligned} |\mathbf{x}-\mathbf{x}'| &= \sqrt{(\mathbf{x}-\mathbf{x}')^2} = \sqrt{\mathbf{x}^2 - 2\mathbf{x}\cdot\mathbf{x}' + \mathbf{x}'^2} \\ &= |\mathbf{x}|\sqrt{1 - 2\frac{\mathbf{x}\cdot\mathbf{x}'}{\mathbf{x}^2} + \frac{\mathbf{x}'^2}{\mathbf{x}^2}} \approx |\mathbf{x}|\left(1 - \frac{\mathbf{x}\cdot\mathbf{x}'}{\mathbf{x}^2}\right) \end{aligned} \tag{2.30}$$

where we neglected $\mathbf{x}'^2/\mathbf{x}^2$ relative to $|\mathbf{x}'|/|\mathbf{x}|$ and used that $\sqrt{1-\varepsilon} \approx 1 - \frac{1}{2}\varepsilon$. So we get

$$k\,|\mathbf{x}-\mathbf{x}'| = k\,|\mathbf{x}| - \left(k\frac{\mathbf{x}}{|\mathbf{x}|}\right)\cdot\mathbf{x}' = k\,|\mathbf{x}| - \mathbf{k}'\cdot\mathbf{x}' \tag{2.31}$$

where we defined

$$\mathbf{k}' = k\,\frac{\mathbf{x}}{|\mathbf{x}|} \tag{2.32}$$

The vector $\mathbf{k}'$ is the momentum of the scattered particle because it is parallel to $\mathbf{x}$ and has magnitude k as it should, see Equation (2.1). Substitution in Equation (2.17) and using $|\mathbf{x}'|/|\mathbf{x}| \ll 1$ gives

$$G(\mathbf{x},\mathbf{x}') = \frac{e^{ik|\mathbf{x}-\mathbf{x}'|}}{|\mathbf{x}|} \approx \frac{e^{ik|\mathbf{x}|}\,e^{-i\mathbf{k}'\cdot\mathbf{x}'}}{|\mathbf{x}|} \tag{2.33}$$

Using this expression in the second term of Equation (2.29) and dropping the remaining terms to implement the Born approximation we get

$$\begin{aligned} \psi(\mathbf{x}) &= e^{i\mathbf{k}\cdot\mathbf{x}} - \left(\frac{2m}{4\pi}\right)\int d^3\mathbf{x}'\,\frac{e^{ik|\mathbf{x}|}e^{-i\mathbf{k}'\cdot\mathbf{x}'}}{|\mathbf{x}|}\,V(\mathbf{x}')\,e^{i\mathbf{k}\mathbf{x}'} \\ &= e^{i\mathbf{k}\cdot\mathbf{x}} - \left(\frac{2m}{4\pi}\right)\int d^3\mathbf{x}'\,e^{i(\mathbf{k}\cdot\mathbf{x}'-\mathbf{k}'\cdot\mathbf{x}')}\,V(\mathbf{x}')\,\frac{e^{ik|\mathbf{x}|}}{|\mathbf{x}|} \end{aligned} \tag{2.34}$$

Comparing this expression with Equation (2.1) we find (with $|\mathbf{x}| = r$) the scattering amplitude in the Born approximation

$$f(\theta,\phi) = -\frac{2m}{4\pi}\int d^3\mathbf{x}\,e^{i(\mathbf{k}-\mathbf{k}')\cdot\mathbf{x}}\,V(\mathbf{x}) \tag{2.35}$$

If we introduce the momentum transfer

$$\mathbf{q} = \mathbf{k} - \mathbf{k}' \tag{2.36}$$

we finally get

$$f(\theta, \phi) = -\frac{2m}{4\pi} \int d^3\mathbf{x}\, e^{i\mathbf{q}\cdot\mathbf{x}}\, V(\mathbf{x}) \tag{2.37}$$

Here, as before, θ and ϕ determine the direction of $\mathbf{k}'$, the momentum of the outgoing particle. The expression $\int d^3\mathbf{x}\, e^{i\mathbf{q}\cdot\mathbf{x}}\, V(\mathbf{x})$ is related to the Fourier transform $\tilde{V}$ of $V(\mathbf{x})$

$$\tilde{V}(\mathbf{q}) = \left(\frac{1}{2\pi}\right)^{\frac{3}{2}} \int d^3\mathbf{x}\, e^{i\mathbf{q}\cdot\mathbf{x}}\, V(\mathbf{x}) \tag{2.38}$$

where the prefactor is conventional (one $\sqrt{2\pi}$ in the denominator for each spatial dimension). Thus the scattering amplitude is proportional to the Fourier transform of the scattering potential. So if an experimentalist measures the scattering cross section as a function of angle then Equation (2.5) gives the scattering amplitude and its inverse Fourier transform gives the scattering potential. Not bad!

We will have more to say about the momentum transfer $\mathbf{q}$ but for now we simply state that the wavelength $\lambda = \hbar/|\mathbf{q}|$ is the one that determines how small a detail of the target (the probed object) we can resolve, as was mentioned at the beginning of Section 2.1.

2.2.5 Electron-Atom Scattering

As an example we will consider electron-atom scattering, a process of great importance presently and historically. Assume that the incident electron is a structureless point particle with an energy E that is much larger than the ionization potential of the atom. Let the atom have atomic number Z, a structureless point charge $Ze\delta^3(\mathbf{x})$ due to the atomic nucleus, and a charge density $\rho(\mathbf{x})$ due to its atomic electrons. Let $\mathbf{x}$ be the coordinate of the incident electron. It will be subject to a potential from the atom

$$V(\mathbf{x}) = -\frac{Ze^2}{|\mathbf{x}|} + Ze^2 \int \frac{\rho(\mathbf{x}')d^3\mathbf{x}'}{|\mathbf{x}-\mathbf{x}'|} = -Ze^2 \int \frac{\rho_t(\mathbf{x}')d^3\mathbf{x}'}{|\mathbf{x}-\mathbf{x}'|} \tag{2.39}$$

The first term represents the potential due to the nucleus, the second term represents the potential from the atomic electrons, and $\rho_t(\mathbf{x})$ is the total charge density of the atom. We have factored out Ze^2 from the integrands so

$$\int \rho(\mathbf{x})\, d^3\mathbf{x} = 1 \quad \text{and} \quad \int \rho_t(\mathbf{x})\, d^3\mathbf{x} = 0 \tag{2.40}$$

We introduce the potential Equation (2.39) in Equation (2.35) for the scattering amplitude

$$
\begin{aligned}
f(\theta,\phi) &= \frac{2m}{4\pi} Ze^2 \int d^3\mathbf{x}\, d^3\mathbf{x}'\, \mathrm{e}^{i\mathbf{q}\cdot\mathbf{x}}\, \frac{\rho_t(\mathbf{x}')}{|\mathbf{x}-\mathbf{x}'|} \\
&= \frac{2m}{4\pi} Ze^2 \int d^3\mathbf{x}'\, \mathrm{e}^{i\mathbf{q}\cdot\mathbf{x}'}\, \rho_t(\mathbf{x}') \int d^3\mathbf{x}\, \frac{\mathrm{e}^{i\mathbf{q}\cdot(\mathbf{x}-\mathbf{x}')}}{|\mathbf{x}-\mathbf{x}'|}
\end{aligned}
\tag{2.41}
$$

where in the second line we have added and subtracted $i\mathbf{q}\cdot\mathbf{x}'$ in the exponential and split the exponential into two pieces such that the integrand of the second integral has $|\mathbf{x}-\mathbf{x}'|$ in its exponential and in its numerator. All $\mathbf{x}$ dependence is in the second integrand and we will do the integral over $\mathbf{x}$, setting $\mathbf{x}-\mathbf{x}' = \mathbf{r}$ and $|\mathbf{x}-\mathbf{x}'| = r$

$$
\mathcal{I} = \int d^3\mathbf{x}\, \frac{\mathrm{e}^{i\mathbf{q}\cdot(\mathbf{x}-\mathbf{x}')}}{|\mathbf{x}-\mathbf{x}'|} = \int d^3\mathbf{x}\, \frac{\mathrm{e}^{i\mathbf{q}\cdot\mathbf{r}}}{r}
\tag{2.42}
$$

The integrand oscillates rapidly for large values of r. To control these oscillations, we introduce a convergence factor $\exp(-\lambda r)$ with $\lambda > 0$ in the integrand and take the limit $\lambda \to 0$ afterward

$$
\mathcal{I} = \int \mathrm{e}^{-\lambda r}\, \frac{\mathrm{e}^{iqr\cos\theta}}{r}\, r^2\, dr\, d\cos\theta\, d\phi
\tag{2.43}
$$

The integral over θ can be done

$$
\int_{-1}^{+1} \mathrm{e}^{iqr\cos\theta}\, d\cos\theta = \frac{\mathrm{e}^{iqr} - \mathrm{e}^{-iqr}}{iqr}
\tag{2.44}
$$

and using it in Equation (2.43) gives

$$
\mathcal{I} = 2\pi \int_0^\infty \mathrm{e}^{-\lambda r} r^2 \left(\frac{\mathrm{e}^{iqr} - \mathrm{e}^{-iqr}}{iqr} \right) dr = \frac{4\pi}{q^2+\lambda^2} \to \frac{4\pi}{q^2}
\tag{2.45}
$$

where in the last step we let $\lambda \to 0$. We leave aside the question of whether it is permitted to exchange the order of the integration and the limit. Using Equation (2.45) in Equation (2.41) we get for the scattering amplitude

$$
f(\theta,\phi) = \frac{2m}{4\pi}\, Ze^2 \frac{4\pi}{q^2} \int d^3\mathbf{x}'\, \mathrm{e}^{i\mathbf{q}\cdot\mathbf{x}'}\, \rho_t(\mathbf{x}')
\tag{2.46}
$$

The total charge density $\rho_t(\mathbf{x})$ is defined by Equation (2.39) and can be written as

$$
\rho_t(\mathbf{x}) = \delta^3(\mathbf{x}) - \rho(\mathbf{x})
\tag{2.47}
$$

Substitution of Equation (2.47) in Equation (2.46) gives

$$f(\theta,\phi) = \frac{2mZe^2}{q^2}\int d^3\mathbf{x}\,e^{i\mathbf{q}\cdot\mathbf{x}}[\delta^3(\mathbf{x}) - \rho(\mathbf{x})] = \frac{2mZe^2}{q^2}[1 - F(\mathbf{q})] \qquad (2.48)$$

with

$$F(\mathbf{q}) = \int d^3\mathbf{x}\,\rho(\mathbf{x})\,e^{i\mathbf{q}\cdot\mathbf{x}} \qquad (2.49)$$

the Fourier transform of the charge density $\rho(\mathbf{x})$ of the atomic electrons. $F(\mathbf{q})$ is called the Form Factor of the atomic electrons. The Form Factor of the nuclear charge, assumed to be a point particle without structure (that is a particle with a charge and mass distribution proportional to $\delta^3(\mathbf{x})$) is apparently 1.

The momentum transfer squared can be calculated as

$$\begin{aligned} q^2 &= \mathbf{q}^2(\mathbf{k} - \mathbf{k}')^2 = (\mathbf{k} - k\mathbf{x}/|\mathbf{x}|)^2 \\ &= k^2 - 2k\,\mathbf{k}\cdot\mathbf{x}/|\mathbf{x}| + k^2 = 2k^2 - 2k^2\cos\theta \\ &= 4k^2\sin^2\tfrac{1}{2}\theta = 8mE\sin^2\tfrac{1}{2}\theta \end{aligned} \qquad (2.50)$$

with θ the scattering angle, that is the angle between $\mathbf{k}$ and $\mathbf{k}'$, or because according to Equation (2.32) $\mathbf{k}'$ is parallel to $\mathbf{x}$, the angle between $\mathbf{k}$ and $\mathbf{x}$. E is the energy of the incident electron and m is the electron mass.

The Form Factor $F(\mathbf{q})$ can be calculated as follows. Let $r = |\mathbf{x}|$ and assume that $\rho(\mathbf{x})$ is spherically symmetric and thus only a function of r. Choose the z-axis along $\mathbf{q}$ so that $\mathbf{q}\cdot\mathbf{x} = qr\cos\chi$ with χ the angle between $\mathbf{q}$ and $\mathbf{x}$. The Form Factor Equation (2.49) can be calculated using spherical coordinates r, χ, ϕ

$$\begin{aligned} F(\mathbf{q}) &= \int r^2 dr\, d\cos\chi\, d\phi\, \rho(r)\, e^{iqr\cos\chi} \\ &= 2\pi\int_0^\infty r^2 dr\,\rho(r)\left[\frac{e^{iqr\cos\chi}}{iqr}\right]_{-1}^{+1} \\ &= 2\pi\int_0^\infty r^2 dr\,\rho(r)\,\frac{e^{iqr} - e^{-iqr}}{iqr} \\ &= \frac{4\pi}{q}\int_0^\infty r\,dr\,\rho(r)\sin qr \end{aligned} \qquad (2.51)$$

To gain further insight into the answer we assume as an example the spherically symmetric Gaussian charge density for the atomic electrons

$$\rho(r) = \frac{1}{(a\sqrt{\pi})^3}\,e^{-(r^2/a^2)} \qquad (2.52)$$

where the constant has been chosen to have $\rho(r)$ normalized to 1; remember that the charge Ze^2 was factored out, see Equation (2.40), and a is a measure of the 'radius' of the charge distribution of the atomic electrons. By substitution of Equation (2.52) in Equation (2.51) we get

$$F(\mathbf{q}) = e^{-(qa/2)^2} \tag{2.53}$$

with $q^2 = \mathbf{q}^2$ given by Equation (2.50). It is seen that the Form Factor is actually a function of q^2, so we write for the Form Factor: $F(q^2)$. Although this charge distribution is not a realistic one for all atoms, it has the important properties that it is localized as atomic electrons are, and that it allows us to do the integral Equation (2.51).

We see from Equation (2.53) that there are two limiting regions: $(qa)^2 \ll 1$ and $(qa)^2 \gg 1$. The region $(qa)^2 \ll 1$ or $q \ll 1/a$ is the region of 'soft' scattering of the incident electron by the atom. Small q^2 means according to Equation (2.50) that the scattering angle θ is small too, and the incident electron will hardly deviate from its original direction (a 'grazing' collision). From Equation (2.53) we see that $F(0)$ is near 1, and from Equation (2.48) that the scattering amplitude $f(\theta, \phi)$ is small. Therefore, with Equation (2.5) the cross section is small and the scattering is correctly called 'soft'. A Form Factor of 1 means that the incident electron sees the atomic electrons as a single point charge as we saw before when we set the charge distribution of the atom's nucleus to $\delta^3(\mathbf{x})$, see Equation (2.47). When q^2 is very small, the nucleus and the atomic electrons both present themselves to the incident electron as point charges located at $\mathbf{x} = 0$, exactly canceling each other's positive and negative charges, and we do not resolve the charge of the nucleus from the charge of the atomic electrons. This is according to our remarks at the beginning of Section 2.1 where we talked about the wavelength of the incident particle and its relation to probing the structure of the target. The wavelength of the probe is $\lambda = 1/q$ and in a soft collision $q \ll 1/a$ so $\lambda \gg a$ so the wavelength of the probe is very large relative to the radius of the target, the form factor equals 1, and we learn little about the internal structure of the target.

On the other hand $(qa)^2 \gg 1$ or $q \gg 1/a$ is the region of 'hard' scattering of the incident electron by the atom. Large q means that according to Equation (2.50) the scattering angle θ is also large (up to π, backward scattering) and the incident electron will deviate substantially from its incident direction. From Equation (2.53) we see that $F(\mathbf{q}) \neq 1$ and that we are sensitive to the functional dependence of F on q^2 and that the scattering amplitude in Equation (2.48) is not small. Therefore, with Equation (2.5) the cross section is not small. By measuring the cross section and inferring the scattering amplitude from it, we can determine the form factor and obtain information about the charge distribution in the target, resolving the nuclear charge from the charge distribution of the atomic electrons, and measuring the charge distribution of the atomic electrons. Clearly, $\lambda \ll a$ here. This is truly probing the structure of the target!

The condition $(qa)^2 = 8mEa^2 \sin^2\theta/2 \gg 1$ for hard scattering can be evaluated as follows. We set a equal to the Bohr radius $a_0 = 1/(Z\alpha m)$ and introduce the ionization potential $E_{\text{ion}} = m(Z\alpha)^2/2$ to get

$$(qa)^2 = 8m\frac{E}{E_{\text{ion}}}E_{\text{ion}}\,a_0^2 = 8m\frac{E}{E_{\text{ion}}}\frac{m}{2}(Z\alpha)^2\left(\frac{1}{Z\alpha m}\right)^2 = 4\frac{E}{E_{\text{ion}}} \tag{2.54}$$

Requiring $(qa)^2 \gg 1$ implies a substantial scattering angle θ and $E/E_{\text{ion}} \gg 1$. The cross section becomes

$$\frac{d\sigma}{d\Omega} = \left(\frac{Ze^2}{4E}\right)^2 \frac{1}{\sin^4\frac{\theta}{2}} \tag{2.55}$$

This is the Rutherford scattering formula, famous for its use in the analysis of the scattering of α particles of atoms, resolving and thus discovering the atomic nucleus. We have used Equation (2.5) to relate the cross section to the scattering amplitude, and Equation (2.48) for the scattering amplitude with $F(\mathbf{q}) = 0$ as we are in the $(qa)^2 \gg 1$ hard scattering domain.

2.3 PHOTO-ELECTRIC EFFECT

The photo-electric effect played a decisive role in the early development of Quantum Physics. Its striking (at the time) characteristics do not have to be discussed here because they are covered in great detail in the early part of any Quantum Physics course. We consider the process

$$\gamma + A \to B + e^- \tag{2.56}$$

where the photon has sufficient energy to ionize the atom. We will model this process as an incident photon ionizing the atom A by ejecting an electron, leaving behind B, the singly ionized version of atom A. We will be more precise later.

Fermi's Golden Rule reads, see Equation (1.103)

$$dw_{fi} = 2\pi \left|\langle B; \mathbf{p}|\lambda V(\mathbf{x})|A; \mathbf{k}, \boldsymbol{\varepsilon}_\lambda(\mathbf{k})\rangle\right|^2 \delta(E_B + E_e - E_A \pm \omega) \tag{2.57}$$

In this expression $\mathbf{p}$ is the momentum of the outgoing electron, $E_e = \mathbf{p}^2/(2m)$ its energy, and $\mathbf{k}$ and λ the momentum and polarization of the incident photon. The $\pm$ sign in the δ-function is the same $\pm$ that is found in the exponent of the harmonic perturbation $H_1(t) = \lambda V(\mathbf{x})\exp(\pm i\omega t)$ and will be resolved shortly. As always, the bra and ket are each products of the appropriate bra and ket of the individual atom, photon and electron. The

Harmonic Perturbation is given by the last two terms of the Hamiltonian Equation (1.77)

$$H_1(t) = -\frac{e}{m}\mathbf{A}\cdot\mathbf{p} + \left(\frac{e}{m}\right)^2 \mathbf{A}^2 \tag{2.58}$$

with **A** given by Equation (1.30)

$$\mathbf{A}(\mathbf{x},t) = \sqrt{\frac{4\pi}{V}}\sum_{\mathbf{k},\lambda}\frac{1}{\sqrt{2\omega_k}}[a_\lambda(\mathbf{k})\,\mathrm{e}^{ikx}\boldsymbol{\varepsilon}_\lambda(\mathbf{k}) + a_\lambda^\dagger(\mathbf{k})\,\mathrm{e}^{-ikx}\boldsymbol{\varepsilon}_\lambda^*(\mathbf{k})] \tag{2.59}$$

We closely follow the procedure used in Section 1.4 to obtain the transition probability. The ket in Equation (2.57) has a photon that is not there in the bra, so we need to select terms in Equation (2.58) that annihilate a single photon. Such terms are present only in the first term of Equation (2.58), while the second term changes the number of photons by 0 and ± 2. The term proportional to $a_\lambda(\mathbf{k})$ in $\mathbf{A}\cdot\mathbf{p}$ is summed over $\mathbf{k}$ but only the term whose values of $\mathbf{k}$ and λ are equal to the ones of the initial photon, listed in the ket, survives. That term has a time dependence $\exp(-i\omega t)$ leading to a $-\omega$ in the δ-function in Equation (2.57). We obtain

$$\begin{aligned}
dw_{fi} &= 2\pi\left|\langle B;\mathbf{p}|\sqrt{\frac{4\pi}{V}}\left(-\frac{e}{m}\right)\frac{1}{\sqrt{2\omega_k}}a_\lambda(\mathbf{k})\,\mathrm{e}^{i\mathbf{k}\cdot\mathbf{x}}\boldsymbol{\varepsilon}_\lambda(\mathbf{k})\cdot\mathbf{p}|A;\mathbf{k},\lambda\rangle\right|^2 \\
&\quad\times\delta(E_B + E_e - E_A - \omega) \\
&= 2\pi\frac{4\pi}{V}\left(\frac{e}{m}\right)^2\frac{1}{2\omega_k}\left|\langle B;\mathbf{p}|\mathrm{e}^{i\mathbf{k}\cdot\mathbf{x}}\boldsymbol{\varepsilon}_\lambda(\mathbf{k})\cdot\mathbf{p}|A\rangle\right|^2\delta(E_B + E_e - E_A - \omega)
\end{aligned} \tag{2.60}$$

Compare this expression with Equation (1.114) for (spontaneous) emission. The only differences are the sign in the exponent of the exponential, the absence of complex conjugation of $\boldsymbol{\varepsilon}_\lambda(\mathbf{k})$, and the sign of ω in the δ-function. The reader should follow the arguments leading to opposite signs for ω in Equation (1.110) and Equation (2.60) to see that conservation of energy comes out automatically. We want to let the operator **p** work backward on the bra $\langle B;\mathbf{p}|$. This can be done because $\boldsymbol{\varepsilon}_\lambda(\mathbf{k})\cdot\mathbf{p}$ commutes with $\exp(i\mathbf{k}\cdot\mathbf{x})$. This can be seen as follows (ψ is a test function)

$$\begin{aligned}
\boldsymbol{\varepsilon}\cdot\mathbf{p}\,\mathrm{e}^{i\mathbf{k}\cdot\mathbf{x}}\,\psi &= \varepsilon_i p_i\,\mathrm{e}^{i\mathbf{k}\cdot\mathbf{x}}\psi \\
&= \varepsilon_i\frac{1}{i}\frac{\partial}{\partial x_i}\mathrm{e}^{i\mathbf{k}\cdot\mathbf{x}}\,\psi \\
&= \varepsilon_i\frac{1}{i}(ik_i)\mathrm{e}^{i\mathbf{k}\cdot\mathbf{x}}\psi + \varepsilon_i\,\mathrm{e}^{i\mathbf{k}\cdot\mathbf{x}}\frac{1}{i}\frac{\partial}{\partial x_i}\psi \\
&= \boldsymbol{\varepsilon}\cdot\mathbf{k}\,\mathrm{e}^{i\mathbf{k}\cdot\mathbf{x}}\psi + \mathrm{e}^{i\mathbf{k}\cdot\mathbf{x}}\boldsymbol{\varepsilon}\cdot\mathbf{p}\,\psi \\
&= \mathrm{e}^{i\mathbf{k}\cdot\mathbf{x}}\boldsymbol{\varepsilon}\cdot\mathbf{p}\,\psi
\end{aligned} \tag{2.61}$$

where we used that $\boldsymbol{\varepsilon} \cdot \mathbf{k} = 0$ in the Coulomb gauge. When we use $\langle B; \mathbf{p}| \boldsymbol{\varepsilon}_\lambda(\mathbf{k}) \cdot \mathbf{p} = \boldsymbol{\varepsilon}_\lambda(\mathbf{k}) \cdot \mathbf{p} \langle B; \mathbf{p}|$ in Equation (2.60) and multiply by the phase space volume $d^3\mathbf{x}\, d^3\mathbf{p}/(2\pi)^3 = V\, d^3\mathbf{p}/(2\pi)^3$ of the outgoing electron, see Equation (1.119), we get

$$dw_{fi} = \frac{4\pi}{(2\pi)^2} \left(\frac{e}{m}\right)^2 \frac{1}{2\omega_k} |\langle B; \mathbf{p}| e^{i\mathbf{k}\cdot\mathbf{x}} |A\rangle|^2 \, [\boldsymbol{\varepsilon}_\lambda(\mathbf{k}) \cdot \mathbf{p}]^2 \, d^3\mathbf{p} \times \delta(E_B + E_e - E_A - \omega) \tag{2.62}$$

Note that V has canceled as it should. Using the δ-function we can integrate over the magnitude of the momentum $|\mathbf{p}|$ of the electron to get

$$dw_{fi} = \frac{4\pi}{8\pi^2} \left(\frac{e}{m}\right)^2 \frac{1}{\omega_k} |\langle B; \mathbf{p}| e^{i\mathbf{k}\cdot\mathbf{x}} |A\rangle|^2 \, [\boldsymbol{\varepsilon}_\lambda(\mathbf{k}) \cdot \mathbf{p}]^2 \, m|\mathbf{p}| \, d\Omega \tag{2.63}$$

where we have used that $E_e = \mathbf{p}^2/(2m)$ and

$$\begin{aligned} & f(\mathbf{p})\, d^3\mathbf{p}\, \delta(E_B + E_e - E_A - \omega_k) \\ &= f(\mathbf{p})\, \delta(E_B + E_e - E_A - \omega_k)\, \mathbf{p}^2\, d|\mathbf{p}|\, d\Omega \\ &= \frac{m}{|\mathbf{p}|} f(\mathbf{p})\, \mathbf{p}^2\, d\Omega \\ &= m\, |\mathbf{p}|\, f(\mathbf{p})\, d\Omega \end{aligned} \tag{2.64}$$

We used that

$$\int f(x)\, \delta[g(x)]\, dx = \sum_i \frac{f(x_i)}{|g'(x_i)|} \tag{2.65}$$

where the x_i are the roots of $g(x) = 0$. The magnitude of $\mathbf{p}$ that makes the argument of the δ-function zero is given by $|\mathbf{p}| = 2m\sqrt{E_B - E_A - \omega_k}$. The bra $\langle B; \mathbf{p}|$ in Equation (2.63) is taken to be the product of the bra $\langle B|$ representing a singly ionized atom B at rest and the bra $\langle \mathbf{p}|$ representing an outgoing electron. It will turn out that the cross section peaks strongly in the forward ($\theta \approx 0$) direction so we represent the outgoing electron by a plane wave $\exp(i\mathbf{p} \cdot \mathbf{x})/\sqrt{V}$ instead of a spherical wave. The ket $|A\rangle$ in Equation (2.63) is taken to be the product of the ket $|A\rangle$ representing the same singly ionized atom at rest and the ket $|e^-\rangle$ representing an electron in a hydrogen-like electron state. The latter can be represented in coordinate space as a $1S$ hydrogen-like wavefunction. We will *not* approximate the exponential $\exp(i\mathbf{k} \cdot \mathbf{x})$ by its first few terms as we did in Section 1.4 on (spontaneous) emission because we do not want to limit the range of validity of the calculation unnecessarily. Thus we can evaluate the matrix element for a hydrogen-like atom with atomic number Z as

$$\langle B; \mathbf{p}|e^{i\mathbf{k}\cdot\mathbf{x}}|A\rangle = \int d^3\mathbf{x}\left(\frac{e^{i\mathbf{p}\cdot\mathbf{x}}}{\sqrt{V}}\right)^* e^{i\mathbf{k}\cdot\mathbf{x}} \frac{1}{\sqrt{\pi}}\left(\frac{Z}{a_0}\right)^{\frac{3}{2}} e^{-\frac{Zr}{a_0}} \tag{2.66}$$

The $\sqrt{V}$ factor in the denominator of the integrand is to normalize the plane wave such that the probability of finding the outgoing electron in the box with volume V is equal to 1. The hydrogen-like wave function is taken to be the one for the ground state

$$\frac{1}{\sqrt{\pi}}\left(\frac{Z}{a_0}\right)^{\frac{3}{2}} e^{-\frac{Zr}{a_0}} \tag{2.67}$$

with a_0 the Bohr radius for hydrogen. The cross section $d\sigma$ is given by Equation (2.7) where $v = 1$ for the incident photon so we get

$$d\sigma = \frac{4\pi}{8\pi^2}\left(\frac{e}{m}\right)^2 \frac{1}{\omega_k}\frac{1}{\pi}\left(\frac{Z}{a_0}\right)^3 [\boldsymbol{\varepsilon}_\lambda(\mathbf{k})\cdot\mathbf{p}]^2 m|\mathbf{p}|\,d\Omega \int d^3\mathbf{x}\, e^{i(\mathbf{k}-\mathbf{p})\cdot\mathbf{x}} e^{-\frac{Z|\mathbf{x}|}{a_0}} \tag{2.68}$$

To do the integral we introduce the momentum transfer $\mathbf{q} = \mathbf{k} - \mathbf{p}$ and $q = |\mathbf{q}|$)

$$\frac{d\sigma}{d\Omega} = 32Z^5 a_0^2 \frac{|\mathbf{p}|}{\omega_k}\left[\frac{\boldsymbol{\varepsilon}_\lambda(\mathbf{k})\cdot\mathbf{p}}{m}\right]^2 \frac{1}{(Z^2 + a_0^2 q^2)^4} \tag{2.69}$$

We evaluate q^2, assuming that $\omega_k \gg E_B - E_A$ (the incident photon's energy is much larger than the ionization potential of the atom but low enough that the outgoing electron is non-relativistic). Then $\omega_k \approx E_e = \mathbf{p}^2/(2m) = \frac{1}{2}mv^2$ (now v is the non-relativistic velocity of the electron) and

$$\begin{aligned} q^2 = \mathbf{q}^2 = (\mathbf{k}-\mathbf{p})^2 &= \mathbf{k}^2 - 2\mathbf{k}\cdot\mathbf{p} + \mathbf{p}^2 \\ &= \omega_k^2 - 2\omega_k|\mathbf{p}|\cos\theta + \mathbf{p}^2 \\ &= \frac{\mathbf{p}^4}{4m^2} - 2\omega_k|\mathbf{p}|\cos\theta + \mathbf{p}^2 \\ &\approx \mathbf{p}^2\,(1 - 2\,\frac{\omega_k}{|\mathbf{p}|}\cos\theta) \\ &= \mathbf{p}^2\,(1 - 2\,\frac{mv^2/2}{mv}\cos\theta) \\ &= \mathbf{p}^2\,(1 - v\cos\theta) \end{aligned} \tag{2.70}$$

where θ is the angle between $\mathbf{k}$ and $\mathbf{p}$ and we neglected in the third line the $\mathbf{p}^4$ term relative to the $\mathbf{p}^2$ term because in the non-relativistic limit $\mathbf{p}^2/(4m^2) \ll \mathbf{p}^2$.

If we assume that the incident photon is unpolarized we must average over the two polarizations by summing over λ and dividing the result by 2. To do the summation over λ we chose the z-axis parallel to $\mathbf{k}$ and $\boldsymbol{\varepsilon}_1(\mathbf{k})$ parallel to the x-axis and $\boldsymbol{\varepsilon}_2(\mathbf{k})$ parallel to the y-axis. Thus $\mathbf{p} = |\mathbf{p}|\,(\sin\theta\cos\phi, \sin\theta\sin\phi, \cos\theta)$, $\boldsymbol{\varepsilon}_1(\mathbf{k}) = (1,0,0)$, and $\boldsymbol{\varepsilon}_2(\mathbf{k}) = (0,1,0)$. We find that

$$\frac{1}{2}\sum_{\lambda=1}^{2}[\boldsymbol{\varepsilon}_\lambda(\mathbf{k})\cdot\mathbf{p}]^2 = \frac{1}{2}(\sin^2\theta\cos^2\phi + \sin^2\theta\sin^2\phi) = \frac{1}{2}\sin^2\theta \tag{2.71}$$

and

$$\frac{d\sigma}{d\Omega} = 2\sqrt{2}\,Z^5\alpha^8 a_0^2\left(\frac{E_e}{m}\right)^{-\frac{7}{2}}\frac{\sin^2\theta}{(1-v\cos\theta)^4} \tag{2.72}$$

This is the desired result for the photo-electric effect. The derivation and the result shows all the manifestations of Quantum Physics. For example, the energy of the outgoing electrons does not increase when one increases the intensity of the incident light. We note that the cross section is not πa_0^2, not even close, because there is a factor α^8 (and more). The cross section is zero for $\theta = 0$ which makes sense because when $\mathbf{p}$ is parallel to $\mathbf{k}$ we have that $\boldsymbol{\varepsilon}_\lambda(\mathbf{k})\cdot\mathbf{p} = 0$ (the $\boldsymbol{\varepsilon}_\lambda(\mathbf{k})$ are perpendicular to $\mathbf{k}$) and the cross section in Equation (2.69) is seen to be proportional to $[\boldsymbol{\varepsilon}_\lambda(\mathbf{k})\cdot\mathbf{p}]^2$. If the energy ω_k of the incident photon is increased, the energy E_e and the velocity v of the outgoing electron both increase, and the cross section starts peaking more and more in the forward direction ($\theta = 0$) and it decreases. At sufficiently high photon energy the cross section will be so small that it will be overtaken by the cross section for $e^+\,e^-$ pair production which starts to be significant when ω_k is well above the threshold of twice the electron mass. Our non-relativistic result in Equation (2.72) is not correct for such high energies and a relativistic calculation is called for.

2.4 PHOTON SCATTERING

2.4.1 Amplitudes

Photon scattering distinguishes itself from the photo-electric effect of the previous section by the fact that there is a photon in the final state as well as in the initial state. Thus we consider

$$\gamma + A \to B + \gamma \tag{2.73}$$

This process is of great importance in condensed matter, biophysics, and other fields of research because, as we shall see, the outgoing photon carries information about the characteristics of the target. We distinguish three cases: elastic scattering with photon energies of the order of the binding energy of the constituents of the target (Rayleigh scattering), elastic scattering with photon energies much larger than the binding energy (Thomson scattering), and inelastic scattering where the incoming and outgoing photons have different energies (Raman scattering). Here, elastic scattering means that the incoming and outgoing photons have equal energy which, because of conservation of energy, implies that A and B stand for an atom in the same state. We shall see that in photon scattering both the $\mathbf{A}\cdot\mathbf{p}$ and $\mathbf{A}^2$ terms in Equation (2.58) contribute, the first one in second order in perturbation theory and the second one in first order. These two contributions are of the same order in e, the electron's charge, and that indicates the need to consider both terms and sum them. The $\mathbf{A}\cdot\mathbf{p}$ term does not contribute in first order because, as we know from our earlier work on spontaneous emission and the photo-electric effect, it changes the number of photons by ± 1. It contributes in second order because, as we will show in the next section, it involves $H_1(t)$ twice, just what is required to have the $\mathbf{A}\cdot\mathbf{p}$ annihilate one photon and create another one.

We start with the calculation to first order with the $\mathbf{A}^2$ term and use Equation (1.94)

$$\lambda\dot{c}_f^{(1)}(t) = -i\langle f|\lambda H_1|n\rangle\, e^{i(E_f-E_i)t} \tag{2.74}$$

where E_f and E_i are the eigenvalues of H_0, the time-independent 'large' part of the Hamiltonian relevant to the atom. The kets $|i\rangle$ for the initial state and for final state $|f\rangle$ are given by

$$|i\rangle = |A;\mathbf{k},\lambda\rangle \qquad |f\rangle = |B;\mathbf{k}',\lambda'\rangle \tag{2.75}$$

and $H_1(t) = [e^2/(2m)]\,\mathbf{A}^2$ so

$$\lambda\dot{c}_f^{(1)}(t) = -i\langle B;\mathbf{k}',\lambda'|\frac{e^2}{2m}\mathbf{A}(\mathbf{x},t)\mathbf{A}(\mathbf{x},t)|A;\mathbf{k},\lambda\rangle\, e^{i(E_B-E_A)t} \tag{2.76}$$

Note that both factors $\mathbf{A}$ are evaluated at the same space-time point. For the matrix element to be non-zero we must pick out from the product $\mathbf{A}(\mathbf{x},t)\,\mathbf{A}(\mathbf{x},t)$ products $a_\lambda(\mathbf{k})a^\dagger_{\lambda'}(\mathbf{k}')$ and $a^\dagger_{\lambda'}(\mathbf{k}')a_\lambda(\mathbf{k})$. These are accompanied by $\exp(-i\omega t)\,\boldsymbol{\varepsilon}_\lambda(\mathbf{k})\exp(+i\omega' t)\,\boldsymbol{\varepsilon}^*_{\lambda'}(\mathbf{k}')$ and $\exp(+i\omega' t)\,\boldsymbol{\varepsilon}^*_{\lambda'}(\mathbf{k}')\exp(-i\omega t)\,\boldsymbol{\varepsilon}_\lambda(\mathbf{k})$ respectively. We get

$$\begin{aligned}\lambda\dot{c}_f^{(1)}(t) &= -i\,\frac{e^2}{2m}\,\frac{4\pi}{2\sqrt{\omega\omega'}}\,\frac{1}{V}\,2\,\boldsymbol{\varepsilon}^*_{\lambda'}(\mathbf{k}')\cdot\boldsymbol{\varepsilon}_\lambda(\mathbf{k})\,e^{i(\omega'-\omega)t}\langle B|A\rangle\, e^{i(E_B-E_A)t}\\ &= -i\,\frac{e^2}{2m}\,\frac{4\pi}{V}\,\frac{1}{\sqrt{\omega\omega'}}\,\delta_{BA}\,\boldsymbol{\varepsilon}^*_{\lambda'}(\mathbf{k}')\cdot\boldsymbol{\varepsilon}_\lambda(\mathbf{k})\,e^{i(E_B-E_A+\omega'-\omega)t}\end{aligned} \tag{2.77}$$

We make the dipole (or long wavelength) approximation $(\mathbf{k}-\mathbf{k}')\cdot\mathbf{x} \ll 1$, see Section 1.4, so we set $\exp[i(\mathbf{k}-\mathbf{k}')\cdot\mathbf{x}] = 1$. The last factor 2 in the first line of Equation (2.77) comes from the fact that the contributions from the two terms discussed above are equal. Because the system A cannot change in elastic scattering we have $\langle B|A\rangle = \delta_{BA}$. The energies E_A and E_B are the eigenvalues of H_0, the Hamiltonian for the systems A and B.The expression is of order e^2.

We now consider the second order contribution of the $\mathbf{A}\cdot\mathbf{p}$ term. We start with the equation for $\dot{c}_f^{(2)}$ Equation (1.92)

$$\lambda^2\dot{c}_f^{(2)}(t) = -i\sum_m \lambda c_m^{(1)}\langle f|\lambda H_1|m\rangle\, e^{i(E_f-E_m)t} \tag{2.78}$$

This second order equation requires a first order expression for $c_m^{(1)}$ for which we use Equation (1.91) where we substitute m for f

$$\lambda\dot{c}_m^{(1)}(t) = -i\sum_n c_n^{(0)}\langle m|\lambda H_1|n\rangle\, e^{i(E_m-E_n)t} \tag{2.79}$$

As in Section 1.3 we assume that the system is in the state $|i\rangle$ at some time t_1 in the past. Then all $c_n^{(0)}(t_1) = 0$ except for $c_i^{(0)}(t_1) = 1$, see the discussion below Equation (1.93) in Section 1.3. We get from Equation (2.79)

$$\lambda\dot{c}_m^{(1)}(t) = -i\langle m|\lambda H_1|i\rangle\, e^{i(E_m-E_i)t} \tag{2.80}$$

We integrate Equation (2.80) from t_1 to t to get

$$\lambda c_m^{(1)}(t) = -i\int_{t_1}^{t} dt'\langle m|\lambda H_1|i\rangle\, e^{i(E_m-E_i)t'} \tag{2.81}$$

where we have introduced a new integration variable t'. Of course $t_1 \leq t' \leq t$. Substitution of Equation (2.81) in Equation (2.78) gives

$$\begin{aligned}\lambda^2\dot{c}_f^{(2)}(t) &= (-i)^2\int_{t_1}^{t} dt'\sum_m \langle f|\lambda H_1|m\rangle\langle m|\lambda H_1|i\rangle\, e^{i(E_f-E_m)t}e^{i(E_m-E_i)t'}\\ &= (-i)^2\left(-\frac{e}{m}\right)^2\int_{t_1}^{t} dt'\sum_I \langle B;\mathbf{k}',\lambda'|\mathbf{A}(\mathbf{x},t)\cdot\mathbf{p}|I\rangle\\ &\quad\langle I|\mathbf{A}(\mathbf{x}',t')\cdot\mathbf{p}|A;\mathbf{k},\lambda\rangle e^{i(E_B-E_I)t}\, e^{i(E_I-E_A)t'}\end{aligned} \tag{2.82}$$

We have replaced m by I where 'I' stands for intermediate. The name is appropriate because the operator product $\mathbf{A}(\mathbf{x}',t')\cdot\mathbf{p}$ converts the state $|A;\mathbf{k},\lambda\rangle$ to the state $|I\rangle$ which in turn is converted by the operator product $\mathbf{A}(\mathbf{x},t)\cdot\mathbf{p}$ to the state $|B;\mathbf{k}',\lambda'\rangle$. Thus the state $|I\rangle$ is truly intermediate; it is only reached in passing from $|A;\mathbf{k},\lambda\rangle$ to $|B;\mathbf{k}',\lambda'\rangle$.

Now consider

$$\langle B;\mathbf{k}',\lambda'|\mathbf{A}(\mathbf{x},t)\cdot\mathbf{p}|I\rangle\langle I|\mathbf{A}(\mathbf{x}',t')\cdot\mathbf{p}|A;\mathbf{k},\lambda\rangle \tag{2.83}$$

The state I can contain either 0 or 2 photons because $\mathbf{A}(\mathbf{x}',t')$ changes the number of photons by ± 1 depending upon which one of the two terms in $\mathbf{A}(\mathbf{x}',t')$ is selected. To get an intermediate state with zero photons, we select from Equation (2.83) $a^{\dagger}_{\lambda'}(\mathbf{k}')\boldsymbol{\varepsilon}^{*}_{\lambda'}(\mathbf{k}')\exp(+i\omega' t)$ from the first matrix element and $a_{\lambda}(\mathbf{k})\boldsymbol{\varepsilon}_{\lambda}(\mathbf{k})\exp(-i\omega t')$ from the second matrix element. To get an intermediate state with two photons, we select $a_{\lambda}(\mathbf{k})\boldsymbol{\varepsilon}_{\lambda}(\mathbf{k})\exp(-i\omega t)$ from the first matrix element and $a^{\dagger}_{\lambda'}(\mathbf{k}')\boldsymbol{\varepsilon}_{\lambda'}(\mathbf{k}')\exp(+i\omega' t')$ from the second matrix element. We use again the dipole (or long wavelength) approximation and set $\exp[(\mathbf{k}-\mathbf{k}')\cdot\mathbf{x}]=1$ and Equation (2.82) becomes

$$\begin{aligned}
\lambda^2\dot{c}^{(2)}_f(t) &= (-i)^2\left(-\frac{e}{m}\right)^2\frac{4\pi}{2V\sqrt{\omega\omega'}}\sum_I\langle B|\boldsymbol{\varepsilon}_{\lambda}(\mathbf{k})\cdot\mathbf{p}\,\mathrm{e}^{-i\omega t}|I\rangle\\
&\quad\times\int_{t_1}^{t}dt'\langle I|\boldsymbol{\varepsilon}^{*}_{\lambda'}(\mathbf{k}')\cdot\mathbf{p}\,\mathrm{e}^{+i\omega' t'}|A\rangle\mathrm{e}^{i(E_B-E_I)t}\mathrm{e}^{i(E_I-E_A)t'}\\
&\quad+(-i)^2\left(-\frac{e}{m}\right)^2\frac{4\pi}{2V\sqrt{\omega\omega'}}\sum_I\langle B|\boldsymbol{\varepsilon}^{*}_{\lambda'}(\mathbf{k}')\cdot\mathbf{p}\,\mathrm{e}^{+i\omega' t}|I\rangle\\
&\quad\times\int_{t_1}^{t}dt'\langle I|\boldsymbol{\varepsilon}_{\lambda}(\mathbf{k})\cdot\mathbf{p}\,\mathrm{e}^{-i\omega t'}|A\rangle\mathrm{e}^{i(E_B-E_I)t}\mathrm{e}^{i(E_I-E_A)t'}
\end{aligned} \tag{2.84}$$

Note the structure of the two terms, it is easy to spot errors when comparing them. We want to integrate this result over t and it is tempting to set all integration limits to $-\infty$ and $+\infty$ as we did in Section 1.3 to get Equation (1.99). This would be incorrect here because the upper integration limit t of the integral over t' is itself an integration variable when integrating $\dot{c}^2_f(t)$. When doing the integral over t', the value of t should be kept fixed at an arbitrary finite value and should not be set to infinity before the integration of t' is completed. So consider the integral over t' in the first term of Equation (2.84)

$$\begin{aligned}
\int_{t_1}^{t}dt'\mathrm{e}^{i(\omega'+E_I-E_A)t'} &= \int_{t_1}^{t}dt'\mathrm{e}^{\lambda t'}\mathrm{e}^{i(\omega'+E_I-E_A)t'}\\
&= \left.\frac{\mathrm{e}^{\lambda t'}\mathrm{e}^{i(\omega'+E_I-E_A)t'}}{\lambda+i(\omega'+E_I-E_A)}\right|_{t_1}^{t}\\
&= \frac{\mathrm{e}^{\lambda t+i(\omega'+E_I-E_A)t}-\mathrm{e}^{\lambda t_1+i(\omega'+E_I-E_A)t_1}}{\lambda+i(\omega'+E_I-E_A)}
\end{aligned} \tag{2.85}$$

In the first line we introduced a convergence factor $\exp(\lambda t')$ with $\lambda > 0$. After the integration we will let $\lambda \to 0$. This procedure damps the rapid oscillations of the integrand when the integration variable $t' \to -\infty$. Mathematicians might have something to say about exchanging integration and taking the limit, but physicists plough ahead and see what happens. When we let $t_1 \to -\infty$ and $\lambda \to 0$, Equation (2.85) becomes

$$\frac{e^{i(\omega'+E_I-E_A)t}}{i(\omega'+E_I-E_A)} \tag{2.86}$$

Likewise we get for the integral over t' in the second term of Equation (2.84)

$$\int_{t_1}^{t} dt' e^{i(-\omega+E_I-E_A)t'} = \frac{e^{i(-\omega+E_I-E_A)t}}{i(-\omega+E_I-E_A)} \tag{2.87}$$

Again, we note the structure of Equation (2.86) and Equation (2.87) when comparing them.

We substitute Equation (2.86) and Equation (2.87) in Equation (2.84) to get

$$\lambda^2 \dot{c}_f^{(2)}(t) = \left(-\frac{e}{m}\right)^2 \frac{4\pi}{2V\sqrt{\omega\omega'}} \frac{1}{i} \sum_I \left[\frac{\langle B|\boldsymbol{\varepsilon}_\lambda(\mathbf{k})\cdot\mathbf{p}|I\rangle\langle I|\boldsymbol{\varepsilon}^*_{\lambda'}(\mathbf{k}')\cdot\mathbf{p}|A\rangle}{E_I - E_A + \omega'} + \frac{\langle B|\boldsymbol{\varepsilon}^*_{\lambda'}(\mathbf{k}')\cdot\mathbf{p}|I\rangle\langle I|\boldsymbol{\varepsilon}_\lambda(\mathbf{k})\cdot\mathbf{p}|A\rangle}{E_I - E_A - \omega} \right] e^{i(E_B+\omega'-E_A-\omega)t} \tag{2.88}$$

This expression is of order e^2, just as Equation (2.77) is. According to Equation (1.85) we must add this result for $\lambda^2 \dot{c}_f^{(2)}(t)$ to the result Equation (2.77) for $\lambda \dot{c}_f^{(1)}(t)$ to get $\dot{c}_f(t)$ to order e^2 (second order in perturbation theory). We then integrate over time from t_1 (at which time the system is in the state $|i\rangle$) to t to get $c_f(t) = \lambda c_f^{(1)}(t) + \lambda^2 c_f^{(2)}(t)$ in the same manner as done with Equation (1.95) in Section 1.3. Remember $c_f^{(0)}(t) = 0$. Note that Equation (2.77) and Equation (2.88) have the same time dependence in their exponentials as they should, because the integration over time will give δ-functions that impose energy conservation and the arguments of the two should therefore be the same. We have the integral, common to Equation (2.77) and Equation (2.88)

$$\int_{t_1}^{t_2} dt\, e^{i(E_B+\omega'-E_A-\omega)t} \tag{2.89}$$

We let $t_1 \to -\infty$ and $t_2 \to +\infty$ and use Equation (1.98) to get

$$\int_{-\infty}^{+\infty} dt\, e^{i(E_B+\omega'-E_A-\omega)t} = 2\pi\, \delta(E_B+\omega'-E_A-\omega) \tag{2.90}$$

Using this in Equation (2.77) and Equation (2.88) and adding the two gives $c_f(\infty)$ using $c_f^{(0)}(\infty) = 0$

$$c_f(\infty) = \lambda c_f^{(1)}(\infty) + \lambda^2 c_f^{(2)}(\infty) = \frac{e^2}{m}\frac{4\pi}{2V\sqrt{\omega\omega'}}\frac{1}{i}\left[\delta_{AB}\,\boldsymbol{\varepsilon}_\lambda(\mathbf{k})\cdot\varepsilon^*_{\lambda'}(\mathbf{k}')\right.$$
$$-\frac{1}{m}\sum_I \frac{\langle B|\boldsymbol{\varepsilon}_\lambda(\mathbf{k})\cdot\mathbf{p}|I\rangle\langle I|\boldsymbol{\varepsilon}^*_{\lambda'}(\mathbf{k}')\cdot\mathbf{p}|A\rangle}{E_I - E_A + \omega'}$$
$$\left.-\frac{1}{m}\sum_I \frac{\langle B|\boldsymbol{\varepsilon}^*_{\lambda'}(\mathbf{k}')\cdot\mathbf{p}|I\rangle\langle I|\boldsymbol{\varepsilon}_\lambda(\mathbf{k})\cdot\mathbf{p}|A\rangle}{E_I - E_A - \omega}\right] 2\pi\delta(E_B + \omega' - E_A - \omega) \tag{2.91}$$

Note the $1/m$ factor in the second and third term. It appears there because we sum the result from the $(e/m)\mathbf{A}$ term ('squared' in second order) and the $(e^2/m)\mathbf{A}^2$ term (in first order) and they have different dependencies upon m.

2.4.2 Cross Section

The transition probability is obtained from $c_f(\infty)$ as explained in Section 1.3. We use Equation (1.100) which contains the square of $c_f(\infty)$ and thus the square of the δ-function for which we use Equation (1.102). To obtain the cross section from the transition probability we use Equation (2.7). We want the cross section for the case that the outgoing photon is in an energy range $d\omega'$ and in a solid angle $d\Omega'$. We therefore multiply the probability and the cross section by the phase space factor in Equation (1.119)

$$\frac{d^3\mathbf{x}\,d^3\mathbf{k}'}{(2\pi)^3} = V\frac{\omega'^2 d\omega' d\Omega'}{(2\pi)^3} \tag{2.92}$$

to get

$$d\sigma = \frac{V}{vT}|c_f(\infty)|^2\, V\frac{\omega'^2 d\omega' d\Omega'}{(2\pi)^3} \tag{2.93}$$

The volume factors V will cancel because $c_f(\infty)$ is proportional to $1/V$ and it gets squared. The time T will cancel because the square of the δ-function is proportional to T. We set the velocity of the incident photon $v = 1$ and introduce the electron radius $r_e = e^2/m$. This is not the 'real' radius but it is obtained by setting the approximate potential energy e^2/r_e of a charge distribution equal to its rest mass energy mc^2. Using

Equation (2.91) in Equation (2.93) and integrating over $d\omega'$ using the δ-function we get

$$\frac{d\sigma}{d\Omega} = r_e^2\left(\frac{\omega'}{\omega}\right)\left| \delta_{AB}\, \boldsymbol{\varepsilon}_\lambda(\mathbf{k}) \cdot \varepsilon^*_{\lambda'}(\mathbf{k}') - \frac{1}{m}\sum_I \frac{\langle B|\boldsymbol{\varepsilon}_\lambda(\mathbf{k}) \cdot \mathbf{p}|I\rangle\langle I|\boldsymbol{\varepsilon}^*_{\lambda'}(\mathbf{k}') \cdot \mathbf{p}|A\rangle}{E_I - E_A + \omega'} \right.$$
$$\left. - \frac{1}{m}\sum_I \frac{\langle B|\boldsymbol{\varepsilon}^*_{\lambda'}(\mathbf{k}') \cdot \mathbf{p}|I\rangle\langle I|\boldsymbol{\varepsilon}_\lambda(\mathbf{k}) \cdot \mathbf{p}|A\rangle}{E_I - E_A - \omega} \right|^2 \tag{2.94}$$

We are now in a position to discuss two (Rayleigh, Thomson) of the three types of scattering mentioned in the beginning of this section because they lend themselves to further development of Equation (2.94).

2.4.3 Rayleigh Scattering

In Rayleigh scattering we are dealing with elastic scattering with photon energies much less than typical energy differences between levels of the target. Thus we have $\omega' = \omega$, $|B\rangle = |A\rangle$, and $\omega = \omega' \ll |E_i - E_A|$. The δ_{AB} term in Equation (2.94) can be evaluated as follows (set $\boldsymbol{\varepsilon}_\lambda(\mathbf{k}) = \boldsymbol{\varepsilon}$ and $\varepsilon^*_{\lambda'}(\mathbf{k}') = \boldsymbol{\varepsilon}'^*$)

$$\delta_{AB}\, \boldsymbol{\varepsilon} \cdot \varepsilon'^* = \varepsilon_i \delta_{ij} \varepsilon_j'^* = i\varepsilon_i\,[p_i, x_j]\,\varepsilon_j'^* = (\boldsymbol{\varepsilon} \cdot \mathbf{p})\,(\boldsymbol{\varepsilon}'^* \cdot \mathbf{x}) - (\boldsymbol{\varepsilon}'^* \cdot \mathbf{x})\,(\boldsymbol{\varepsilon} \cdot \mathbf{p}) \tag{2.95}$$

where we have used Equation (1.7). Sandwiching this between $\langle A|$ and $|A\rangle$ we get

$$\langle A|\boldsymbol{\varepsilon} \cdot \varepsilon'^*|A\rangle = i\langle A|(\boldsymbol{\varepsilon} \cdot \mathbf{p})\,(\boldsymbol{\varepsilon}'^* \cdot \mathbf{x}) - (\boldsymbol{\varepsilon}'^* \cdot \mathbf{x})\,(\boldsymbol{\varepsilon} \cdot \mathbf{p})|A\rangle$$
$$= i\sum_I \left[\langle A|\boldsymbol{\varepsilon} \cdot \mathbf{p}|I\rangle\langle I|\boldsymbol{\varepsilon}'^* \cdot \mathbf{x}|A\rangle - \langle A|\boldsymbol{\varepsilon}'^* \cdot \mathbf{x}|I\rangle\langle I|\boldsymbol{\varepsilon} \cdot \mathbf{p}|A\rangle\right] \tag{2.96}$$

where we have twice inserted the completeness relation $\sum_I |I\rangle\langle I| = 1$. We now use Equation (1.115) to relate the matrix elements of the operators $\mathbf{p}$ and $\mathbf{x}$ and replace the two occurrances of $\mathbf{p}$ in Equation (2.96) by $\mathbf{x}$.

$$\langle A|\boldsymbol{\varepsilon} \cdot \mathbf{p}|I\rangle = mi(E_A - E_I)\langle I|\boldsymbol{\varepsilon} \cdot \mathbf{x}|A\rangle \tag{2.97}$$

Thus we get from Equation (2.96)

$$\langle A|\boldsymbol{\varepsilon} \cdot \varepsilon'^*|A\rangle = i\sum_I im(E_A - E_I)\langle A|\boldsymbol{\varepsilon} \cdot \mathbf{x}|I\rangle\langle I|\boldsymbol{\varepsilon}'^* \cdot \mathbf{x}|A\rangle$$
$$- i\sum_I im(E_I - E_A)\langle A|\boldsymbol{\varepsilon}'^* \cdot \mathbf{x}|I\rangle\langle I|\boldsymbol{\varepsilon} \cdot \mathbf{x}|A\rangle \tag{2.98}$$

Making the same replacement of **p** by **x** in the second and third terms in Equation (2.94) and substitution of Equation (2.98) in Equation (2.94) with $\omega' = \omega$ gives

$$\frac{d\sigma}{d\Omega} = r_e^2 m^2 \Bigg| \sum_I (E_I - E_A)\langle A|\boldsymbol{\varepsilon}\cdot\mathbf{x}|I\rangle\langle I|\boldsymbol{\varepsilon}'^*\cdot\mathbf{x}|A\rangle$$
$$+ \sum_I (E_I - E_A)\langle A|\boldsymbol{\varepsilon}'^*\cdot\mathbf{x}|I\rangle\langle I|\boldsymbol{\varepsilon}\cdot\mathbf{x}|A\rangle$$
$$- \sum_I (E_I - E_A)^2 \frac{\langle A|\boldsymbol{\varepsilon}\cdot\mathbf{x}|I\rangle\langle I|\boldsymbol{\varepsilon}'^*\cdot\mathbf{x}|A\rangle}{E_I - E_A + \omega}$$
$$- \sum_I (E_I - E_A)^2 \frac{\langle A|\boldsymbol{\varepsilon}'^*\cdot\mathbf{x}|I\rangle}{\langle I|\boldsymbol{\varepsilon}\cdot\mathbf{x}|A\rangle} E_I - E_A - \omega \Bigg|^2 \qquad (2.99)$$

where we have written all energy differences as $E_I - E_A$, introducing minus signs as needed. Note the sign of each term. In Rayleigh scattering $\omega \ll |E_A - E_I|$ so we have neglected the ω terms in the denominators of the third and fourth term and they cancel the squares in those terms. Inspection of Equation (2.99) shows that the first and third terms as well as the second and fourth terms sum to zero, so the cross section approaches zero in the limt $\omega \to 0$. This indicates that we have substantial cancellations among the four terms and we proceed as follows. The first and third terms in Equation (2.99) are both proportional to $\langle A|\boldsymbol{\varepsilon}\cdot\mathbf{x}|I\rangle\langle I|\boldsymbol{\varepsilon}'^*\cdot\mathbf{x}|A\rangle$ so we sum their prefactors (taking the common factor m outside the modulus symbols)

$$(E_I - E_A) - \frac{(E_I - E_A)^2}{E_I - E_A + \omega} = (E_I - E_A) - \frac{E_I - E_A}{1 + \frac{\omega}{E_I - E_A}}$$
$$= (E_I - E_A) - (E_I - E_A)\left[1 - \frac{\omega}{E_I - E_A} + \left(\frac{\omega}{E_I - E_A}\right)^2 + \ldots\right]$$
$$= \omega - \frac{\omega^2}{E_I - E_A} + \ldots \qquad (2.100)$$

where we have expanded the denominator in the last term of the first line using

$$\frac{1}{1+x} = 1 - x + x^2 \ldots \qquad (2.101)$$

The terms with $(E_I - E_A)$ in Equation (2.100) cancel, consistent with our earlier observation. The second and fourth terms in Equation (2.99) are both proportional to $\langle A|\boldsymbol{\varepsilon}'^*\cdot\mathbf{x}|I\rangle\langle I|\boldsymbol{\varepsilon}\cdot\mathbf{x}|A\rangle$ so we sum their prefactors and obtain a relation similar to Equation (2.100) but with $\omega \to -\omega$. Substituting Equation (2.100) and its equivalent for $\omega \to -\omega$ in Equation (2.99) and watching the signs carefully, we discover that terms proportional to ω cancel too (use $\sum_I |I\rangle\langle I| = 1$) leaving terms proportional to ω^2. We obtain

$$\frac{d\sigma}{d\Omega} = r_e^2 m^2 \omega^4 \left| \sum_I \frac{\langle A|\boldsymbol{\varepsilon}\cdot\mathbf{x}|I\rangle\langle I|\boldsymbol{\varepsilon}'^*\cdot\mathbf{x}|A\rangle + \langle A|\boldsymbol{\varepsilon}'^*\cdot\mathbf{x}|I\rangle\langle I|\boldsymbol{\varepsilon}\cdot\mathbf{x}|A\rangle}{E_I - E_A} \right|^2 \tag{2.102}$$

The units of the right and left hand side of Equation (2.102) are length squared, so we do not need to insert powers of $\hbar$ or c. Note that we must sum over all possible intermediate states I of the system A and we must do that before we square. Note too that the matrix elements in the terms are proportional to the (transition) electric dipole moments of the system A, a consequence of our dipole (or long wavelength) approximation. The cross section is proportional to ω^4 and thus inversely proportional to λ^4 due to the cancellation of the terms proportional to ω discussed earlier. Had this cancellation not happened, the cross section would have been proportional to ω^2 and thus inversely proportional to λ^2. For atoms in colorless gases the energy difference $|E_I - E_A|$ corresponds to the ultraviolet (wavelengths less than blue light) region so the approximation $\omega \ll |E_I - E_A|$ is valid. The λ dependence of the elastic scattering of light was discovered by Rayleigh using classical electromagnetism with wavelengths in the optical region. The absence of $\hbar$ in Equation (2.102) is the reason that Rayleigh was able to derive the formula without the use of Quantum Physics. The strong dependence upon wavelength is responsible for the blue sky and red sunsets. Short wavelength (blue) light gets scattered much more than long wavelength (red) light. So if you look at a part of the sky away from the sun, light must scatter off molecules in the atmosphere to reach your eyes and mostly blue light will reach them. Looking at the sunset the opposite happens. Scattering will preferentially remove blue light leaving the red to reach your eyes.

2.4.4 Thomson Scattering

In Thomson scattering we are dealing with elastic scattering with photon energies much larger than typical energy differences between levels in the target. Thus we have $\omega' = \omega \gg |E_I - E_A|$. Because $E_I - E_A$ appear in the denominators of the second and third terms of Equation (2.94), the first term with δ_{AB} dominates and we have

$$\frac{d\sigma}{d\Omega} = r_e^2 \left|\boldsymbol{\varepsilon}_\lambda(\mathbf{k}) \cdot \boldsymbol{\varepsilon}^*_{\lambda'}(\mathbf{k}')\right|^2 \tag{2.103}$$

Note that elastic scattering requires $|\mathbf{k}| = |\mathbf{k}'|$ but *not* $\lambda = \lambda'$. The cross section depends upon the angle between $\boldsymbol{\varepsilon}_\lambda(\mathbf{k})$ and $\boldsymbol{\varepsilon}^*_{\lambda'}(\mathbf{k}')$ and not directly upon the angle between $\mathbf{k}$ and $\mathbf{k}'$ (the scattering angle). This makes for interesting results. To evaluate Equation (2.103) we must specify $\boldsymbol{\varepsilon}_\lambda(\mathbf{k})$ and $\boldsymbol{\varepsilon}^*_{\lambda'}(\mathbf{k}')$, keeping in mind that they are perpendicular to the respective photon momenta $\mathbf{k}$ and $\mathbf{k}'$. We choose the z-axis along $\mathbf{k}$, the direction of the incident photon. Therefore we choose $\boldsymbol{\varepsilon}_1(\mathbf{k})$ along the x-axis and $\boldsymbol{\varepsilon}_2(\mathbf{k})$ along the y-axis and obtain a right-handed triplet of unit vectors $\boldsymbol{\varepsilon}_1(\mathbf{k}), \boldsymbol{\varepsilon}_2(\mathbf{k}), \mathbf{k}/\omega_k$. We characterize $\mathbf{k}'$ in spherical coordinates as $\mathbf{k}' = \omega_k(\sin\theta\cos\phi, \sin\theta\sin\phi, \cos\phi)$. We choose $\boldsymbol{\varepsilon}_1(\mathbf{k}')$ to be real and to be in the x,y-plane perpendicular to $\mathbf{k}'$. Thus $\boldsymbol{\varepsilon}_1(\mathbf{k}') = (a, b, 0)$ with a and b to be determined such that $a^2 + b^2 = 1$. The requirement that $\boldsymbol{\varepsilon}_1(\mathbf{k}')$ is perpendicular to $\mathbf{k}'$ gives $a\omega\sin\theta\cos\phi + b\omega\sin\theta\sin\phi = 0$. These two relations between a and b give $a = \sin\phi$ and $b = -\cos\phi$ or $a = -\sin\phi$ and $b = \cos\phi$. We choose the first solution so $\boldsymbol{\varepsilon}_1(\mathbf{k}') = (\sin\phi, -\cos\phi, 0)$, independent of θ. We define $\boldsymbol{\varepsilon}_2(\mathbf{k}')$ to be real as well and given by $\boldsymbol{\varepsilon}_2(\mathbf{k}') = (\mathbf{k}'/\omega') \times \boldsymbol{\varepsilon}'_1(\mathbf{k}')$ where the order of the vectors in the cross product has been chosen such that we obtain a right-handed triplet of unit vectors $\boldsymbol{\varepsilon}_1(\mathbf{k}'), \boldsymbol{\varepsilon}_2(\mathbf{k}'), \mathbf{k}'/\omega_k$. We find that $\boldsymbol{\varepsilon}_2(\mathbf{k}') = (\cos\theta\cos\phi, \cos\theta\sin\phi, -\sin\theta)$. With these expressions for the polarization vectors we find

$$\begin{aligned}
\boldsymbol{\varepsilon}_1(\mathbf{k}) \cdot \boldsymbol{\varepsilon}_1(\mathbf{k}') &= \sin\phi & \boldsymbol{\varepsilon}_2(\mathbf{k}) \cdot \boldsymbol{\varepsilon}_1(\mathbf{k}') &= -\cos\phi \\
\boldsymbol{\varepsilon}_1(\mathbf{k}) \cdot \boldsymbol{\varepsilon}_2(\mathbf{k}') &= \cos\theta\cos\phi & \boldsymbol{\varepsilon}_2(\mathbf{k}) \cdot \boldsymbol{\varepsilon}_2(\mathbf{k}') &= \cos\theta\sin\phi
\end{aligned} \tag{2.104}$$

If we assume that the incident photon is unpolarized we must average over its two possible polarizations, that is sum over λ and divide by 2. If we assume that the polarization of the outgoing photon is not measured we must sum over its two possible polarizations, that is sum over λ'. Note that we must calculate the sum of squared terms and not the square of the sum. We get

$$\begin{aligned}
\frac{d\sigma}{d\Omega} &= r_e^2 \frac{1}{2} \sum_{\lambda,\lambda'=1}^{2} \left|\boldsymbol{\varepsilon}_\lambda(\mathbf{k}) \cdot \boldsymbol{\varepsilon}_{\lambda'}(\mathbf{k}')\right|^2 \\
&= \tfrac{1}{2} r_e^2 \left(\sin^2\phi + \cos^2\theta\cos^2\phi + \cos^2\phi + \cos^2\theta\sin^2\phi\right) \\
&= \tfrac{1}{2} r_e^2 \left(1 + \cos^2\theta\right)
\end{aligned} \tag{2.105}$$

This result exhibits an interesting angular dependence of the cross section: it is maximum for $\theta = 0$ or π, that is, for outgoing photons along the incident photon's direction or in the opposite direction. It is minimal at directions perpendicular to the incident photon's direction. There is no ϕ dependence as expected because there is no direction defined transverse to the incident photon's direction for unpolarized incident photons. The total cross section is obtained by integrating Equation (2.106) over θ

$$\sigma = \frac{8\pi}{3} r_e^2 \tag{2.106}$$

This is called the Thomson cross section, and it is a factor 8/3 larger than the 'area' πr_e^2 of the electron. The units in Equation (2.106) are correct as is, there is no need to add powers of $\hbar$ or c. As was the case for Rayleigh scattering, this allowed Thomson to derive Equation (2.106) using classical electrodynamics.

If the incident photon is polarized along the x-axis but the outgoing photon's polarization is not observed we get

$$\begin{aligned} \frac{d\sigma}{d\Omega} &= r_e^2 \sum_{\lambda'=1}^{2} \left|\boldsymbol{\varepsilon}_1(\mathbf{k}) \cdot \boldsymbol{\varepsilon}_{\lambda'}(\mathbf{k}')\right|^2 \\ &= r_e^2(\sin^2\phi + \cos^2\theta\cos^2\phi) \end{aligned} \tag{2.107}$$

Now there is ϕ dependence, made possible by the fact that a direction perpendicular to the incident photon's direction is defined by its polarization. Interestingly, for forward scattering ($\theta = 0$) we find that the ϕ dependence disappears as it should.

If the incident photon is unpolarized and the outgoing photon is required to be polarized along $\boldsymbol{\varepsilon}'_1(\mathbf{k}')$ we get

$$\begin{aligned} \frac{d\sigma}{d\Omega} &= r_e^2 \tfrac{1}{2} \textstyle\sum_{\lambda=1}^{2} \left|\boldsymbol{\varepsilon}_\lambda(\mathbf{k}) \cdot \boldsymbol{\varepsilon}_1(\mathbf{k}')\right|^2 \\ &= \tfrac{1}{2} r_e^2(\sin^2\phi + \cos^2\phi) \\ &= \tfrac{1}{2} r_e^2 \end{aligned} \tag{2.108}$$

This cross section is independent of θ and ϕ. If instead we require the outgoing photon to be polarized along $\boldsymbol{\varepsilon}'_2(\mathbf{k}')$ we get

$$\begin{aligned} \frac{d\sigma}{d\Omega} &= r_e^2 \tfrac{1}{2} \textstyle\sum_{\lambda=1}^{2} \left|\boldsymbol{\varepsilon}_\lambda(\mathbf{k}) \cdot \boldsymbol{\varepsilon}_2(\mathbf{k}')\right|^2 \\ &= \tfrac{1}{2} r_e^2(\cos^2\theta\cos^2\phi + \cos^2\theta\sin^2\phi) \\ &= \tfrac{1}{2} r_e^2 \cos^2\theta \end{aligned} \tag{2.109}$$

This is a remarkable result when compared with Equation (2.108). It is zero for $\theta = \pi/2$ and equals $\frac{1}{2} r_e^2$ for $\theta = 0$ or π. We see that scattering at a certain angle can cause the outgoing photon to be polarized and the polarization is a maximum for $\theta = \pi/2$. Reflect on why the roles of $\boldsymbol{\varepsilon}_1(\mathbf{k}')$ and $\boldsymbol{\varepsilon}_2(\mathbf{k}')$ are so different.

PROBLEMS

(1) *Scattering in the Born Approximation.*
Consider the scattering of a particle by an infinitely massive target that presents a spherically symmetric potential $V(r)$ to the projectile. $V(r)$ is given by $V(r) = V_0$ for $r < a$ and $V(r) = 0$ for $r > a$. Use the Born approximation in your calculations.
(a) Calculate the scattering amplitude $f(\theta, \phi)$ where θ, ϕ are the scattering angles.
(b) Calculate the cross section differential in the scattering angles.
(c) Derive the condition for the Born approximation to be valid.
(d) How does the validity of the Born approximation depend upon the energy of the incident particle and the magnitude of V_0?
(e) Study the limiting case where the potential is a δ-function by taking the limit $a \to 0$ while keeping the product aV_0 constant.

(2) *Neutron Scattering.*
Low energy neutron scattering is an important tool of research in condensed matter. Consider the scattering of a very low energy neutron of an electron. The electron resides in an atom that is embedded in the lattice of a crystal. The energy of the neutron is so low that it may be assumed that the electron is undisturbed by the scatter. The vector potential at the position $\mathbf{r}_e$ of the electron, due to the magnetic moment $\boldsymbol{\mu}$ of the neutron at position $\mathbf{r}$, is given by

$$\mathbf{A} = \boldsymbol{\mu} \times \frac{(\mathbf{r}_e - \mathbf{r})}{(\mathbf{r}_e - \mathbf{r})^3} \tag{2.110}$$

This vector potential leads to an interaction term in the Hamiltonian of the form

$$V(\mathbf{x}) = -\frac{e}{2m}\left[\mathbf{p} \cdot \mathbf{A}(\mathbf{x}) + \mathbf{A}(\mathbf{x}) \cdot \mathbf{p}\right] - \frac{e}{2m}\boldsymbol{\sigma} \cdot \nabla \times \mathbf{A}(\mathbf{x}) \tag{2.111}$$

with e and m the electron's charge and mass respectively, $\mathbf{p}$ the electron's momentum, $\boldsymbol{\sigma}$ its Pauli spin operator, and $\mathbf{x} = \mathbf{r}_e - \mathbf{r}$.
(a) Derive the equation for $V(\mathbf{x})$.
(b) Prove that

$$\int d^3\mathbf{x}\, e^{-i\mathbf{q}\cdot\mathbf{x}} \mathbf{A}(\mathbf{x}) = -\frac{4\pi i}{\mathbf{q}^2} \boldsymbol{\mu} \times \mathbf{q} \tag{2.112}$$

(c) Show that

$$\langle f | V(\mathbf{x}) | i \rangle = \frac{4\pi i e}{m} \frac{1}{\mathbf{q}^2} \boldsymbol{\mu} \times \mathbf{q} \cdot (e^{i\mathbf{q}\cdot\mathbf{r}_e}\mathbf{p}) - \frac{4\pi e}{2m}\left[\boldsymbol{\sigma} \cdot \boldsymbol{\mu} - \frac{(\boldsymbol{\sigma} \cdot \mathbf{q})(\boldsymbol{\mu} \cdot \mathbf{q})}{\mathbf{q}^2}\right] e^{i\mathbf{q}\cdot\mathbf{r}_e} \tag{2.113}$$

in the Born approximation. Here $|i\rangle$ and $|f\rangle$ represent the incoming and outgoing neutron and $\mathbf{q} = \mathbf{k} - \mathbf{k}'$ with $\mathbf{k}$ and $\mathbf{k}'$ the momenta of the incoming and outgoing neutron.

Consider the long wavelength limit $\mathbf{q}^2 \to 0$ in each of the two terms in (c) and average over $\mathbf{q}$.

(d) Show that the first term averages to

$$\frac{8\pi}{3}\mu_B\,\boldsymbol{\mu}\cdot\mathbf{L} \tag{2.114}$$

with $\mathbf{L}$ the angular momentum of the electron.

(e) Show that the second term averages to

$$\frac{8\pi}{3}\mu_B\,\boldsymbol{\mu}\cdot\boldsymbol{\sigma} \tag{2.115}$$

(f) Comment on the results.

(3) *Form Factor.*

Consider the spherically symmetric charge density $\rho(\mathbf{r}) = A\exp(-\mu r)$.

(a) Find the value of A that normalizes the charge density.

(b) Calculate the Form Factor that corresponds to this charge density.

3

Symmetries and Conservation Laws

3.1 SYMMETRIES AND CONSERVATION LAWS

3.1.1 Symmetries

Symmetries and conservations laws are intimately connected, in classical physics as well as in quantum physics. To study this connection we introduce the notion of a symmetry operation: an operation on a physical system is a symmetry operation if it leaves the intrinsic properties of the physical system invariant. We say that the physical system has a symmetry. Symmetry operations have associated with them symmetry *operators* that work on the ket associated with the physical system: an operator U is called a symmetry operator if the ket $|\psi\rangle$ and the ket $|\psi'\rangle = U|\psi\rangle$ represent the same intrinsic properties of the physical system. In other words, the intrinsic properties of the system are invariant under the operation of a symmetry operator on the system's ket. Or in still other words: the sets of eigenvalues of all commuting operators that correspond to the intrinsic properties of the physical system are invariant under the operation of a symmetry operator.

There are two kinds of operators (and thus two kinds of symmetry operators): continuous ones and discrete ones. Continuous operators depend upon one or more continuous variables while a discrete operator does not. Examples of continuous operators are the three translation operators T_x, T_y and T_z (they correspond to a shift of the origin of a coordinate system in the x, y and z directions respectively), the three rotation operators R_x, R_y and R_z (they correspond to a rotation of a coordinate system around the x, y

and z axes respectively), the operator that changes the phase of all kets, and many more. Examples of discrete operators are the parity operator P (which we have encountered in Section 1.6), the charge conjugation operator C (it corresponds to a change of particles into anti-particles and vice versa), the time reversal operator T (it corresponds to a change of t to $-t$), the permutation operator $\mathcal{P}_{i,j}$ (it corresponds to an exchange of the labels i and j of two particles), translations and rotations of a coordinate system by a fixed amount (useful for example in the study of crystals), and many more. We include the word *intrinsic* in our definitions because, for example, the translation of the origin of a coordinate system will obviously change the eigenvalue of the position operator X, but it leaves other intrinsic properties such as energy, momentum, angular momentum, charge, mass, parity, and so on unchanged, and it is the latter property that is required for the translation to be a symmetry operation.

We can phrase a symmetry property in still another way: a symmetry of a physical system has associated with it an unobservable quantity. For example, symmetry under translations implies that the *absolute* position of a physical system is unobservable (but its position *relative* to a coordinate system is observable, see above). This coordinate system can be chosen at will, and the properties of a physical system do not depend upon that choice. We summarize this by saying thatspace is uniform. Similarly, symmetry under rotations implies that the absolute orientation of a physical system is unobservable. Thus the direction of the axes of a coordinate system can be chosen at will and we say that space is isotropic. There is a deep connection with the Cosmological Principle that forms the basis of Cosmology. This principle states that space is uniform and isotropic. If absolute velocity is unobservable we conclude that the intrinsic properties of a physical system are invariant under Lorenz transformations. Our inability to tell the difference between inertial mass and gravitational mass (the Equivalence Principle) leads to the theory of General Relativity.

3.1.2 Conservation Laws

The connection between a symmetry of a physical system and a conservation law can be demonstrated as follows. If the state of a physical system is represented by the ket $|\psi\rangle$ then the system's energy E is given by

$$H|\psi\rangle = E|\psi\rangle \tag{3.1}$$

The operator U connects the kets $|\psi\rangle$ and $|\psi'\rangle$

$$|\psi'\rangle = U|\psi\rangle \tag{3.2}$$

If we apply this operator U to Equation (3.1) from the left, we obtain

$$UH|\psi\rangle = UE|\psi\rangle = E\,U|\psi\rangle \tag{3.3}$$

If U is indeed a symmetry operator then the energy associated with the ket $|\psi'\rangle$ must be the same as the energy associated with the ket $|\psi\rangle$ so

$$H|\psi'\rangle = E|\psi'\rangle \quad \text{or} \quad HU|\psi\rangle = EU|\psi\rangle \tag{3.4}$$

Comparing Equation (3.3) and Equation (3.4), we conclude that

$$UH = HU \quad \text{or} \quad [U, H] = 0 \tag{3.5}$$

In general, the time dependence of an operator A that has no explicit time dependence itself is given by its commutation relation with the Hamiltonian

$$\frac{dA}{dt} = i[H, A] \tag{3.6}$$

When we apply this relation to the symmetry operator U, we obtain with Equation (3.5)

$$\frac{dU}{dt} = 0 \tag{3.7}$$

Operators that commute share the same eigenkets (or in the case of degeneracy, shared eigenkets can be constructed). If U is a symmetry operator, U and H commute according to Equation (3.5) and U and H share the same eigenkets (or in the case of degeneracy, shared eigenkets can be constructed). Thus the eigenvalues of U and the energy E of the physical system are observable simultaneously. Alternatively, the product of the uncertainties in the eigenvalues of U and H is proportional to the expectation value of the commutator of U and H. Because this commutator is zero, there is no uncertainty relation for the eigenvalues of U and H. Because of Equation (3.7), the eigenvalues of U are independent of time, that is they are conserved. Thus we have established a deep connection between a symmetry of a physical system and a conservation law. The argument can be reversed: if a physical system is observed to obey a conservation law, it must have a symmetry associated with it.

Operators form a Group. Group Theory is a branch of mathematics that is eminently suitable to discuss symmetry operators, but it is overkill to develop its formalism just for the present application to Quantum Physics. Instead we will obtain results in a straightforward but less elegant manner.

3.2 CONTINUOUS SYMMETRY OPERATORS

3.2.1 Translations

We take as an example the translation operator that corresponds to a translation of a coordinate system along its x-axis by an amount a_x in the positive x-direction. Label the original coordinate system A and label the translated coordinate system B. A physical system located at a position x as measured in coordinate system B can be represented by the ket $|x\rangle$. The system will appear to be at the position $x + a_x$ in the coordinate system A, and it can be represented by the ket $|x + a_x\rangle$. We define the operator T_x that connects the two kets as in Equation (3.2)

$$|x + a_x\rangle = T_x|x\rangle \tag{3.8}$$

where the kets are eigenkets of the position operator X

$$X|x\rangle = x|x\rangle \quad \text{and} \quad X|x + a_x\rangle = (x + a_x)|x + a\rangle \tag{3.9}$$

The subscript x on T_x reminds us that the translation is along the x-direction. We require that the kets $|x + a_x\rangle$ and $|x\rangle$ have the same norm, that is $\langle x + a_x|x + a_x\rangle = \langle x|x\rangle$ or $\langle x|T^\dagger T|x\rangle = \langle x|x\rangle$. Because this relation must hold independently of the choice of the kets $|x\rangle$ we have

$$T_x^\dagger T_x = 1 \tag{3.10}$$

The inverse operator of T_x is T_x^{-1} and satisfies by definition the relation $T_x^{-1}T_x = T_x T_x^{-1} = 1$. Using Equation (3.10) we find

$$T_x^{-1} = T_x^\dagger \tag{3.11}$$

so T_x is a unitary operator. Note that T_x is not Hermitian so its eigenvalues, if any, are not real and thus are not observable.

Applying T_x^{-1} from the left to Equation (3.8) we obtain

$$T_x^{-1}|x + a_x\rangle = |x\rangle \tag{3.12}$$

This shows that the operator T_x^{-1} corresponds to a translation of a coordinate system by an amount a_x in the negative x-direction, as expected.

We will now derive an expression for the operator T. Consider the combination $T_x X T_x^{-1}$ working on the ket $|x + a_x\rangle$

$$\begin{aligned} T_x X T_x^{-1}|x + a_x\rangle &= T_x X|x\rangle = T_x x|x\rangle = x T_x|x\rangle = x|x + a_x\rangle \\ &= (X - a_x 1)|x + a_x\rangle \end{aligned} \tag{3.13}$$

where we used Equation (3.12) for the first equality, Equation (3.9) for the second equality, Equation (3.8) for the fourth equality, and Equation (3.9) for the last equality. This relation must hold for any ket $|x + a_x\rangle$ so we have the operator equation

$$T_x X T_x^{-1} = X - a_x 1 \tag{3.14}$$

We define exp(A), the exponentiation of the operator A, by its Taylor series expansion

$$e^A = 1 + A + \tfrac{1}{2}A^2 + \tfrac{1}{6}A^3 + \cdots \tag{3.15}$$

It follows from this definition that the inverse operator of exp(A) is exp($-A$). Without loss of generality we can write the operator T_x in the form of an exponential

$$T_x = e^{-ia_x P_x} \tag{3.16}$$

where P_x is an operator yet to be determined. The usefulness of writing T_x in the form Equation (3.16) will become clear soon. For the moment we note that Equation (3.16) gives $T_x = 1$ for $a_x = 0$, as expected because if $a_x = 0$ the kets $|x + a_x\rangle$ and $|x\rangle$ should be equal. The imaginary unit i in the exponent in Equation (3.16) together with the fact that T_x is unitary, see Equation (3.11), means that the operator P_x is Hermitian (show this) and that its eigenvalues, if any, will be real and observable. This remedies the fact that the eigenvalues of T_x are not observable.

To identify the operator P_x in Equation (3.16) we consider the case of an infinitely small a_x. We expand Equation (3.16) in powers of a_x and substitute the result in Equation (3.14). Keeping terms to order a^1 we get $(1 - ia_x P_x)X(1 + ia_x P^\dagger) = X - a_x 1$ or $iP_x X - iXP_x = 1$ or $[P_x, X] = 1/i$. We conclude that the unknown operator P_x introduced in Equation (3.16) is the x-component of the momentum operator **P**. Had we not introduced the minus sign in Equation (3.16) we would have concluded that the unknown operator P_x was the negative of the momentum operator and we would be led to introduce the minus sign.

If T_x is a symmetry operator it must commute with the Hamiltonian, see Equation (3.5). It is easy to show using Equation (3.15) that if T_x commutes with H that P_x also commutes with H. Therefore P_x and H share the same eigenkets (or in the case of degeneracy, shared eigenkets can be constructed). Thus the eigenvalues of P_x and the energy E of the physical system are simultaneously observable and the eigenvalues of P_x are conserved. Thus we have derived conservation of the x component of the momentum **P** from the symmetry of the physical system under translations in the x direction.

In general we may write a continuous unitary operator U as

$$U = e^{-i\varepsilon G} \tag{3.17}$$

G is called the generator of U, and G is Hermitian thanks to the presence of the imaginary unit in the exponent. We have shown above that P_x is the generator for the translation operator T_x in the limit that a is infinitely small. Translations over a finite distance a_x in the x-direction can be obtained by an infinitely large number n of infinitely small translations δa_x in the x-direction where δa_x satisfies the relation $n\delta a_x = a_x$. Because the translation operator for one translation over an infinitely small distance δa_x in the x-direction is given by $1 - i\delta a_x P_x$, the translation operator for n such translations is given by $(1 - i\delta a_x P_x)^n = [1 - i(a_x/n)P_x]^n$. When we take the limit $n \to \infty$ and use the definition

$$\lim_{n\to\infty}\left(1 + \frac{x}{n}\right)^n = \mathrm{e}^x \tag{3.18}$$

we find that the translation operator for translations over a finite distance a takes the form of Equation (3.16).

Translations along the y and z directions are generated by P_y and P_z respectively. The operator for a translation by an arbitrary distance $\mathbf{a}$ can be written as

$$T = T_x T_y T_z = \mathrm{e}^{i\mathbf{P}\cdot\mathbf{a}} \tag{3.19}$$

Because P_x, P_y and P_z commute among themselves, the order of the factors in Equation (3.19) does not matter. This can be verified using the definition in Equation (3.15) of an exponentiated operator.

3.2.2 Rotations

This is not the case for rotations. The operator that corresponds to a rotation of a coordinate system around the z-axis by an amount ϕ is

$$R_z(\phi) = \mathrm{e}^{-i\phi L_z} \tag{3.20}$$

analogous to translations. Here L_z, the generator of rotations around the z-axis, is the z-component of the orbital angular momentum. To show this, use that $L_z = (1/i)\partial/\partial\phi$. If R_z is a symmetry operator it must commute with the Hamiltonian, see Equation (3.5). As before, if R_z commutes with H then L_z also commutes with H. Therefore the eigenvalues of L_z are conserved. We have derived the conservation of the z component of angular momentum from the symmetry of the physical system under rotations around the z axis.

Rotations around the x and y axes are generated by L_x and L_y respectively. Care has to be exercised when writing a general rotation as a product of several rotations around different axes, because the operators L_x, L_y and L_z do not commute among themselves. Of course L_z commutes with itself and thus the operator that corresponds to two successive rotations ϕ_1 and ϕ_2 around the z axis can be written as the product of the two individual operators $R_z(\phi_1)$ and $R_z(\phi_2)$ corresponding to each rotation. It follows from Equation (3.20) that the operator product for two successive rotations over angles ϕ_1 and ϕ_2 is the same as the single operator for a rotation over an angle equal to the sum of ϕ_1 and ϕ_2. Operators generated by commuting generators form an Abelian group, while operators generated by non-commuting generators form a non-Abelian group. The latter category of operators is extremely important in the theory of the unification of the electromagnetic and weak interactions in elementary particle physics.

With the deep connection between translational and rotational symmetries and momentum and angular momentum conservation respectively, we have found a connection between the Cosmological Principle mentioned earlier and conservation of momentum and angular momentum.

3.3 DISCRETE SYMMETRY OPERATORS

We briefly review discrete operators; briefly because we have discussed an example of a discrete symmetry operator, the parity operator P, in Section 1.6. We take the permutation operator as an example. Consider a physical system consisting of not necessarily equivalent particles. We label the particles with labels $1, 2, \ldots$. If the particles are different we can do this unambiguously by, for example, giving the most massive of the particles the label 1, the next most massive particle label 2, and so on. The permutation operator $\mathcal{P}_{i,j}$ is defined as the operator that exchanges the labels i and j

$$\mathcal{P}_{i,j}\left|1, 2, \cdots i, \cdots j, \cdots\right\rangle \to \left|1, 2, \cdots j, \cdots i, \cdots\right\rangle \tag{3.21}$$

To keep the norm of the transformed ket $\left|1, 2, \cdots j, \cdots i, \cdots\right\rangle$ the same as the norm of the original ket $\left|1, 2, \cdots i, \cdots j, \cdots\right\rangle$ we require, as in the case of continuous operators, that $\mathcal{P}_{i,j}$ is a unitary operator

$$\mathcal{P}_{i,j}^{\dagger} = \mathcal{P}_{i,j}^{-1} \tag{3.22}$$

Contrary to the case with continuous operators, the discrete operators can be shown to be Hermitian if their square equals the unit operator. This is the case for $\mathcal{P}_{i,j}$, P and C, but not for the rotation operators that correspond to a rotation over a finite angle (unless that angle is π). The permutation operator satisfies the relation

$$\mathcal{P}_{i,j}\mathcal{P}_{i,j} = 1 \tag{3.23}$$

By definition

$$\mathcal{P}_{i,j}^{-1}\mathcal{P}_{i,j} = \mathcal{P}_{i,j}\mathcal{P}_{i,j}^{-1} = 1 \tag{3.24}$$

Comparing this relation with Equation (3.23) we find that

$$\mathcal{P}_{i,j}^{-1} = \mathcal{P}_{i,j} \tag{3.25}$$

as expected. Comparing this relation with Equation (3.22) we find that

$$\mathcal{P}_{i,j} = \mathcal{P}_{i,j}^{\dagger} \tag{3.26}$$

so $\mathcal{P}_{i,j}$ is Hermitian and its eigenvalues (if any) are real and observable.

If the two particles are identical we cannot define in an unambiguous way how they are to be labeled, and either choice is as good as the other because the system's intrinsic properties cannot depend upon the choice. Therefore, exchanging the labels i and j of two identical particles is a symmetry operation and thus $\mathcal{P}_{1,2}$ commutes with the Hamiltonian, see the discussion leading to Equation (3.5). As with the continuous symmetry operators, $\mathcal{P}_{i,j}$ and H share the same eigenkets (or in case of degeneracy, shared eigenkets can be constructed). Thus the eigenvalues of $\mathcal{P}_{i,j}$ and the energy E of the physical system are simultaneously observable and the eigenvalues of $\mathcal{P}_{i,j}$ are conserved. It can be seen from Equation (3.23) that the eigenvalues $\eta_{\mathcal{P}}$ of $\mathcal{P}_{i,j}$ satisfy

$$\eta_{\mathcal{P}}^2 = 1 \tag{3.27}$$

so that

$$\eta_{\mathcal{P}} = \pm 1 \tag{3.28}$$

This means that the eigenkets of H are either even (symmetric) or odd (anti-symmetric) under the permutation of the labels i and j. Because the eigenvalues $\eta_{\mathcal{P}}$ are conserved, such symmetry is true at all times.

Examples of such physical systems are assemblies of fermions or of bosons. Systems consisting of identical fermions are represented by a ket that is anti-symmetric at all times under exchange of two labels. Systems consisting of identical bosons are represented by a ket that is symmetric at all times under exchange of two labels. From this it follows that two or more identical fermions can not share the same quantum numbers, because such a ket is identically zero. This is of course the Pauli exclusion principle. It forms the basis of the shell structure found in atomic and nuclear physics.

The above discussion was for the exchange operator $\mathcal{P}_{i,j}$. It also holds, with obvious modifications, for the parity operator P, for the charge conjugation operator C, and for all other discrete operators whose square is equal to the unit operator. They have important applications in elementary particle physics where they lead to conservation laws.

The time reversal operator requires a separate discussion because it changes kets into bras and bras into kets. Its treatment is outside the scope of this book; it is discussed in any course on elementary particle physics.

3.4 DEGENERACY

3.4.1 Example

Degeneracy in quantum physics means that an operator has two or more eigenkets that all share the same eigenvalue. Two kinds of degeneracy can be distinguished: accidental and non-accidental degeneracy. The latter is due to a symmetry of the physical system, while the former is what the term says: accidental and not due to a symmetry.

The elementary treatment of the hydrogen atom with the Schrödinger equation provides an example of the two different kinds of degeneracy. States with principal quantum number n have energies

$$E(n) = -\frac{\alpha^2 m}{2n^2} \tag{3.29}$$

with α the fine structure constant and m the electron mass. The angular momentum ℓ can take the values $0, 1, \cdots, n-1$. The energy $E(n)$ does not depend upon ℓ and thus a state with principal quantum number n is n-fold degenerate. This degeneracy is due to the $1/r$ dependence of the potential energy term in the Hamiltonian. Any other r-dependence of the potential energy would make the energy dependent upon ℓ and thus lift (remove) the degeneracy. No symmetry is involved.

The z-component of the angular momentum $\mathbf{L}$ can take the values $m = -\ell, -\ell+1, \cdots, +\ell$. The energy $E(n)$ in Equation (3.29) does not depend upon the value of m either, and thus a state with quantum numbers n, ℓ is $(2\ell+1)$-fold degenerate. It can be shown that this degeneracy is due to the rotational symmetry of the system. If we break the rotational symmetry by defining a preferred direction in space, for example with a static external electric or magnetic field, the degeneracy will be lifted. A hydrogen atom with quantum numbers n, ℓ and m in a static external magnetic field B has an energy equal to

$$E(n, \ell_z) = -\frac{\alpha^2 m}{2n^2} + \frac{eB}{2m} m \qquad (m = \ell, \ell+1, \cdots, \ell) \tag{3.30}$$

and the $(2\ell+1)$-fold degenerate energy level $E(n, \ell)$ splits into $2\ell+1$ distinct energy levels. If we let the magnetic field go to zero and rotational symmetry is restored, we see from Equation (3.30) that the degeneracy is restored as well.

The role of symmetry in degeneracy is as follows. If a physical system has a symmetry, the operator U associated with it commutes with the Hamiltonian H, see the arguments leading up to Equation (3.5). Therefore, U and H share eigenkets (or in the case of degeneracy, shared eigenkets can be constructed). An eigenket of H with eigenvalue E can therefore carry additional labels, namely the eigenvalues of U in the case of a discrete symmetry operator, or in the case of a continuous symmetry operator, the eigenvalues of its generator. If the symmetry operator U has m eigenvalues for the eigenket of H with energy E, an m-fold degeneracy *may* result depending upon the detailed properties of the system. If the symmetry is broken, as in the example above, U is not a symmetry operator anymore, U and H do not commute anymore, and there are in general no eigenkets that are shared between U and H. The degeneracy is lifted.

3.4.2 Isospin

As often is the case in art, broken symmetries in physics are sometimes more interesting. There are numerous examples of this in atomic, molecular, condensed matter, nuclear, and elementary particle physics. For example, the proton and neutron have nearly the same mass: $m_p = 938.27\,\text{MeV}$ and $m_n = 939.57\,\text{MeV}$. The difference is $1.30\,\text{MeV}$. Heisenberg considered the hypothesis that the proton and the neutron are manifestations of a single particle, the nucleon, and that the nucleon N can be represented by either one of two kets, the ket $|p\rangle$ for the proton and the ket $|n\rangle$ for the neutron. Both kets are eigenkets of H_{N}, the Hamiltonian of the nuclear interaction, and m_{N} is the single energy belonging to the two kets

$$H_{\text{N}}\,|p\rangle = m_{\text{N}}\,|p\rangle \qquad H_{\text{N}}\,|n\rangle = m_{\text{N}}\,|n\rangle \tag{3.31}$$

It is said that the proton and the neutron form a doublet.

In reality, the two-fold degeneracy is lifted but only slightly, supposedly by the electromagnetic interaction H_{em}. In this example, the electromagnetic interaction plays the role of the static magnetic field in the example in the previous subsection. We can imagine that we can turn off the electromagnetic interaction, like we could turn off the static magnetic field earlier, in order to restore the symmetry and the two-fold degeneracy of the system. If we do that, the question arises as to what the symmetry is that causes the resulting two-fold degeneracy, and what the symmetry operator is that commutes with H_{N} and whose eigenvalues can be used to label the degenerate levels. Heisenberg assumed it to be a rotational symmetry (compare with the example in the previous subsection) in 'isospin space' that leaves the system, described by H_{N} alone, invariant. Isospin space is an imaginary three-dimensional space where the three components I_x, I_y and I_z of the 'isospin' operator **I** are plotted along each of the three axes respectively. The isospin operator **I** is the generator of rotations in isospin space in analogy with **L** being the generator of rotations in ordinary space. The three components of **I** satisfy therefore the same commutation relations as the angular momentum operator **L**. Only I^2 and one of the three components of **I** can be defined simultaneously, analogous to the situation with **L**. We chose I^2 and I_z. We assign $I = \frac{1}{2}$ to the nucleon and $I_z = +\frac{1}{2}$ to the proton and $I_z = -\frac{1}{2}$ to the neutron. Rotations in isospin space are assumed to be symmetry operations, I^2 and I_z commute with H_N so they share eigenkets, and the eigenvalues of I^2 and I_z are conserved. The two degenerate levels are labeled by I_z. Turning on the electromagnetic interaction breaks the rotational symmetry in isospin space and the two-fold degeneracy of the nucleon is lifted. This makes sense because the proton is charged and the neutron is neutral, so the electromagnetic interaction will act differently on the two.

This formalism seems contrived but it has practical consequences in processes where the nuclear interaction described by H_N playes a role: we have conservation of isospin (seperately for I^2 and I_z) and we have selection rules that specify whether a process is allowed or not by the nuclear interaction. The selection rules are experimentally verified up to a precision limited by the electromagnetic interaction that causes a violation of the selection rules. This violation is small because H_{em} is much smaller than H_N, witness the small mass difference between the proton and the neutron.

It is interesting that we are able to derive two new selection rules without knowledge of the form of the Hamiltonian H_N, using only symmetry arguments. Indeed, the form of H_N is not known.

The nucleon is not the only application of the isospin formalism. There are numerous examples in elementary particle physics of multiplets of particles that form degenerate states in the absence of H_{em}. A famous example is the triplet π^+, π^0 and π^- with isospin 1. Other values of isospin than $\frac{1}{2}$ and 1 are encountered in elementary particle physics.

PROBLEMS

(1) *Quantum Numbers.*
Explain under what conditions quantum numbers are additive or multiplicative.

(2) Consider the Hamiltonian of the hydrogen atom. Do not assume that the proton is infinitely heavy at rest.
  (a) Are the translation operators symmetry operators?
  (b) Same questions for the rotation operators.
  (c) Discuss the consequences of (a) and (b).
  (d) Show that the Hamiltonan is invariant under charge conjugation. What is a consequence?

4

Relativistic Quantum Physics

4.1 KLEIN-GORDON EQUATION

When Schrödinger wrote down his equation, he was well aware that it was a non-relativistic one because he used the non-relativistic Hamiltonian $H = \mathbf{p}^2/(2m) + V$ as a starting point. Apparently he first tried to use the Einstein relation

$$E^2 - \mathbf{p}^2 = m^2 \tag{4.1}$$

and made the substitution in Equation (1.5)

$$\mathbf{p} \to \frac{1}{i}\nabla \qquad H\,(\text{or } E) \to -\frac{1}{i}\frac{\partial}{\partial t} \tag{4.2}$$

to obtain

$$\left(\nabla^2 - \frac{\partial^2}{\partial t^2}\right)\phi(\mathbf{x}, t) - m^2\phi(\mathbf{x}, t) = 0 \tag{4.3}$$

where we have introduced the wavefunction $\phi(\mathbf{x}, t) = \phi(x)$. Using the equivalent substitution in Equation (1.6) or rewriting Equation (4.3), this can be written more compactly as

$$\left(\partial^2 - m^2\right)\phi(x) = 0 \tag{4.4}$$

Schrödinger rejected this equation because he was unable to define a probability density $\rho(x)$ that was positive definite while his non-relativistic equation leads naturally to a probability density $\rho(x) = \psi^*\psi$ which is obviously positive definite. This problem was resolved in 1934 by Pauli and Weiskopf by introducing second quantization of Equation (4.4) in much the same way as we introduced second quantization of the electromagnetic field in Section 1.2. Negative probability density was interpreted as the probability density of an anti-particle. The Equation (4.3) was reintroduced and called the Klein-Gordon equation. We show the derivation of the probability and probability current for the Klein-Gordon equation. The argument closely follows the derivation of the same quantities for the non-relativistic Schrödinger equation.

We start with Equation (4.3) and multiply it from the left by ϕ^* to obtain

$$\phi^* \nabla^2\phi - \phi^* \frac{\partial^2\phi}{\partial t^2} - m^2\phi^*\phi = 0 \tag{4.5}$$

Next we take the complex conjugate of Equation (4.3) and multiply it from the right by ϕ to get

$$(\nabla^2\phi^*)\phi - \frac{\partial^2\phi^*}{\partial t^2}\phi - m^2\phi^*\phi = 0 \tag{4.6}$$

Note that the notation $(\nabla^2\phi^*)\phi$ means that the operator ∇^2 works on ϕ^* only. We subtract Equation (4.6) from Equation (4.5) to get

$$\left[\phi^* \nabla^2\phi - (\nabla^2\phi^*)\phi\right] - \left(\phi^* \frac{\partial^2\phi}{\partial t^2} - \frac{\partial^2\phi^*}{\partial t^2}\phi\right) = 0 \tag{4.7}$$

or

$$\nabla \cdot \left[\phi^* \nabla\phi - (\nabla\phi^*)\phi\right] - \frac{\partial}{\partial t}\left(\phi^* \frac{\partial\phi}{\partial t} - \frac{\partial\phi^*}{\partial t}\phi\right) = 0 \tag{4.8}$$

In analogy with the definition of the probability current for the non-relativistic Schrödinger equation, we define the probability current as

$$\mathrm{j} = \frac{1}{2im}\left(\phi^* \nabla\phi - (\nabla\phi^*)\phi\right) \tag{4.9}$$

To obtain a Lorenz four-vector $j_\mu = (\mathrm{j}, i\rho)$ for the probability current and probability density we must define the probability density ρ as

$$i\rho = \frac{1}{2im}\left(\phi^* \partial_4\phi - (\partial_4\phi^*)\phi\right) \tag{4.10}$$

or with $\partial_4 = \partial/\partial(it)$

$$\rho = -\frac{1}{2im}\left(\phi^* \frac{\partial\phi}{\partial t} - \frac{\partial\phi^*}{\partial t}\phi\right) \tag{4.11}$$

To obtain Equation (4.10) from Equation (4.9) we replaced the i-th component of the spatial derivatives $\nabla_i = \partial_i$ by ∂_4 in Equation (4.9). Substitution of Equation (4.9) and Equation (4.11) into Equation (4.8) gives the law of conservation of probability

$$\nabla \cdot \mathbf{j} - \frac{\partial \rho}{\partial t} = 0 \tag{4.12}$$

or

$$\partial_\mu j_\mu = 0 \tag{4.13}$$

similar to the one obtained for the non-relativistic Schrödinger equation. One can use Gauss' theorem to show that Equation (4.13) implies the conservation of probability.

The problem encountered by Schrödinger can now be seen: the probability density Equation (4.11) is not positive definite. Dirac solved this problem in 1928 by writing an entirely new wave equation, discussed in the next section. Second quantization of the Klein-Gordon Equation (4.4) gave a natural interpretation to positive and negative probability densities. After second quantization the wavefunction becomes an operator that corresponds to particle and anti-particle creation and annihilation. Positive and negative values of the probability density simply correspond to particles and anti-particles and j_μ corresponds to the flow of a probability density current. The field ϕ carries no indices (compare with the electromagnetic field A_μ) so it describes particles with spin 0 (as opposed to the electromagnetic field A_μ wich carries spin 1). So the Klein-Gordon equation is suitable for the description of spin 0 particles, also called scalar particles. Note that A_μ and ϕ satisfy the same equation.

4.2 DIRAC EQUATION

4.2.1 Derivation of the Dirac Equation

Before Pauli-Weiskopf did the second quantization of the Klein-Gordon equation in 1934, Dirac attempted to resolve the problem of negative probability density by reducing the Klein-Gordon equation to a set of four differential equations that are first order in the derivative with respect to time. He was motivated by the realization that the possibly negative probability density was caused by the second derivative in time in the Klein-Gordon Equation (4.3). After all, the Schrödinger equation does not suffer from a negative probability density. When tracing the derivation of non-relativistic probability density using the Schrödinger equation, is it seen that the first derivative with respect to time is the determining factor. But Lorenz invariance requires that the derivatives with respect to time and space are of the same order (first order in this case).

The lefthand side of Equation (4.3) contains second-order derivatives in space and time. We bring in Equation (4.3) the m^2 term to the right-hand side and take the square root to obtain

$$\sqrt{\left(\nabla^2 - \frac{\partial^2}{\partial t^2}\right)}\,\phi(\mathbf{x},t) = m\,\phi(\mathbf{x},t) \tag{4.14}$$

With $x_4 = it$ we can write the left-hand side of Equation (4.14) as

$$\sqrt{\frac{\partial^2}{\partial x_1^2} + \frac{\partial^2}{\partial x_2^2} + \frac{\partial^2}{\partial x_3^2} + \frac{\partial^2}{\partial x_4^2}} = m \tag{4.15}$$

If x and y are numbers, $\sqrt{x^2 + y^2} \neq x + y$ but we may try to set

$$\sqrt{x^2 + y^2} = ax + by \tag{4.16}$$

The a and b are objects of an as yet unknown nature that do not necessarily commute with each other. We find the conditions on a and b by squaring Equation (4.16) to get

$$x^2 + y^2 = a^2x^2 + (ab + ba)\,xy + b^2y^2 \tag{4.17}$$

where we have taken care not to commute the objects a and b. The numbers x and y are assumed to commute with the objects a and b and of course with each other. Equation (4.17) is satisfied if the objects a and b satisfy

$$a^2 = 1 \qquad ab + ba = \{a, b\} = 0 \qquad b^2 = 1 \tag{4.18}$$

Here $\{a, b\}$ is the anti-commutator of a and b. Numbers cannot satisfy the anti-commutator relation in Equation (4.18) but matrices can. Examples of matrices with properties required in Equation (4.18) are the Pauli matrices

$$\begin{pmatrix} 0 & 1 \\ 1 & 0 \end{pmatrix} \qquad \begin{pmatrix} 0 & -i \\ i & 0 \end{pmatrix} \qquad \begin{pmatrix} 1 & 0 \\ 0 & -1 \end{pmatrix} \tag{4.19}$$

which satisfy

$$\{\sigma_i, \sigma_j\} = 2\delta_{ij} \tag{4.20}$$

where δ_{ij} is the Kronecker delta symbol. So for example $a = \sigma_1$ and $b = \sigma_2$ satisfy the three conditions in Equation (4.18).

Following Dirac, we now generalize this to the square root of the sum of the squares of four terms and set

$$\sqrt{\frac{\partial^2}{\partial x_1^2} + \frac{\partial^2}{\partial x_2^2} + \frac{\partial^2}{\partial x_3^2} + \frac{\partial^2}{\partial x_4^2}} = -\left(\gamma_1\frac{\partial}{\partial x_1} + \gamma_2\frac{\partial}{\partial x_2} + \gamma_3\frac{\partial}{\partial x_3} + \gamma_4\frac{\partial}{\partial x_4}\right) \tag{4.21}$$

To compare with Equation (4.16), the minus sign in Equation (4.21) is conventional and merely redefines the γ_μ. To find the conditions on the objects γ_μ we follow the example above and square Equation (4.21) to find the conditions

$$\gamma_\mu^2 = 1 \text{ for } \mu = 1,2,3,4 \qquad \gamma_\mu \gamma_\nu + \gamma_\nu \gamma_\mu = 0 \text{ for } \mu \neq \nu \tag{4.22}$$

The conditions on the γ_μ in Equation (4.22) can be written in a compact notation as

$$\gamma_\mu \gamma_\nu + \gamma_\nu \gamma_\mu = \{\gamma_\mu, \gamma_\nu\} = 2\,\delta_{\mu\nu} \tag{4.23}$$

As was the case earlier, numbers cannot satisfy the anti-commutation relation in Equation (4.23) but matrices can, so we anticipate that the γ will be matrices yet to be determined. Substitution of Equation (4.21) in Equation (4.15) and multiplying the resulting operator equation from the right by ψ we obtain

$$\left[\gamma_1 \frac{\partial}{\partial x_1} + \gamma_2 \frac{\partial}{\partial x_2} + \gamma_3 \frac{\partial}{\partial x_3} + \gamma_4 \frac{\partial}{\partial x_4}\right]\psi + m\,\psi = 0 \tag{4.24}$$

or

$$(\gamma_\mu \partial_\mu + m)\,\psi = 0 \tag{4.25}$$

where as always, repeated indices are summed over. Our convention is that repeated Greek indices such as μ are summed over from 1 to 4 while repeated Latin indices such as i are summed over from 1 to 3. The combination $\gamma_\mu a_\mu$, with a_μ an arbitrary four-vector, appears often so we introduce the shorthand

$$\not{a} = \gamma_\mu a_\mu \tag{4.26}$$

so that Equation (4.26) becomes

$$(\not{\partial} + m)\,\psi = 0 \tag{4.27}$$

This is called the Dirac equation. It is manifestly Lorenz covariant.

We shall see that this seemingly simple equation describes particles and anti-particles with spin $\frac{1}{2}$. These particles have been observed, confirming the validity of the Dirac equation. When we introduce the electromagnetic interaction into the Dirac equation, it will turn out to assign a magnetic moment with a g-factor $g = 2$ to the Dirac particles. This factor 2 is the value that was introduced in an *ad hoc* manner in the discussion of a spin $\frac{1}{2}$ particle in an external magnetic field.

To obtain the Hamiltonian that corresponds to the Dirac Equation (4.27) we rewrite it showing explicitly its time and spatial components

$$(\gamma_i \partial_i + \gamma_4 \partial_4 + m)\,\psi = 0 \tag{4.28}$$

Now multiply Equation (4.28) from the left by γ_4 to get

$$(\partial_4 + \gamma_4\gamma_i\partial_i + \gamma_4 m)\,\psi = 0 \tag{4.29}$$

When we make the inverse of the replacement Equation (4.2) and replace $E \to H$ we get

$$H = i\gamma_4\gamma_i p_i + \gamma_4 m \tag{4.30}$$

$$= \alpha_i p_i + \beta m \tag{4.31}$$

$$= \boldsymbol{\alpha} \cdot \mathbf{p} + \beta m \tag{4.32}$$

where we defined

$$\alpha_i = i\gamma_4\gamma_i \quad \text{and} \quad \beta = \gamma_4 \tag{4.33}$$

The 'vector' $\boldsymbol{\alpha}$ has components $\alpha_1, \alpha_2, \alpha_3$ and is 'dotted' into $\mathbf{p}$ in Equation (4.32). The equation is written in a way that is easy to remember. Because we anticipate the γ_μ to be matrices, the α_i and β are expected to be matrices as well. The 'vector' $\boldsymbol{\alpha}$ has matrices as its 'components'. It takes a little getting used to!

For the Hamiltonian Equation (4.32) to be Hermitian we require that the α_i and β matrices be Hermitian

$$\alpha_i^\dagger = \alpha_i \qquad \beta^\dagger = \beta \tag{4.34}$$

The second relation implies that

$$\gamma_4 = \gamma_4^\dagger \tag{4.35}$$

while the first relation implies

$$(i\gamma_4\gamma_i)^\dagger = i\gamma_4\gamma_i \tag{4.36}$$

or

$$-i\gamma_i^\dagger\gamma_4^\dagger = -i\gamma_i\gamma_4 \tag{4.37}$$

On the left side we have used that the Hermitian conjugate of a product of two matrices is equal to the product of their complex conjugates but in reverse order. On the right side we used that γ_4 and γ_i $(i = 1, 2, 3)$ anti-commute, see Equation (4.23). We see from Equation (4.37) that

$$\gamma_i^\dagger = \gamma_i \tag{4.38}$$

We can combine Equation (4.35) and Equation (4.38) into

$$\gamma_\mu^\dagger = \gamma_\mu \tag{4.39}$$

Thus for the Hamiltonian to be Hermitian, all four γ matrices must be Hermitian as well. Because $\gamma_\mu^2 = 1$ (no summation over μ), see Equation (4.23), it follows that $\gamma_\mu^{-1} = \gamma_\mu$ so the γ matrices equal their respective inverses. Because the γ matrices are Hermitian, see Equation (4.39), we have

$$\gamma_\mu^{-1} = \gamma_\mu = \gamma_\mu^\dagger \tag{4.40}$$

so the γ matrices are unitary and their determinants are ± 1. The latter property can also be derived from the fact that $\gamma_\mu^2 = 1$ so $\det(\gamma_\mu^2) = 1$ or

$$\det \gamma_\mu = \pm 1 \tag{4.41}$$

The γ matrices must be traceless (have zero trace) because $\mathrm{Tr}(\gamma_\nu \gamma_\mu \gamma_\nu)$ ($\nu \neq \mu$, no summation over ν) can be evaluated as

$$\mathrm{Tr}(\gamma_\nu \gamma_\mu \gamma_\nu) = -\mathrm{Tr}(\gamma_\nu \gamma_\nu \gamma_\mu) \tag{4.42}$$

$$= -\mathrm{Tr}\, \gamma_\mu \tag{4.43}$$

where in Equation (4.42) we used that γ_ν and γ_μ anti-commute if $\nu \neq \mu$ and that $\gamma_\nu \gamma_\nu = 1$ (no summation over ν). But because the trace of a product of matrices does not change when the matrices are cyclically permuted, we also have (do a right-handed cyclic permutation)

$$\mathrm{Tr}(\gamma_\nu \gamma_\mu \gamma_\nu) = \mathrm{Tr}(\gamma_\mu \gamma_\nu \gamma_\nu) \tag{4.44}$$

$$= \mathrm{Tr}(\gamma_\mu) \tag{4.45}$$

Comparing Equation (4.42) and Equation (4.45) we find

$$\mathrm{Tr}\, \gamma_\mu = 0 \ (\mu = 1, 2, 3, 4) \tag{4.46}$$

These many properties of the γ matrices are important when we next determine their explicit representation.

The four γ_μ cannot be matrices of second or third order. One can see this as follows. We know from Equation (4.23) that $\gamma_\mu \gamma_\nu = -\gamma_\nu \gamma_\mu$ if $\mu \neq \nu$. Taking the determinant on both sides we get $\det(\gamma_\mu \gamma_\nu) = \det(-\gamma_\nu \gamma_\mu)$ or

$$\det \gamma_\mu \det \gamma_\nu = \det \gamma_\nu \det(-\gamma_\mu) \tag{4.47}$$

Because $\det \gamma_\nu \neq 0$, see Equation (4.41), we conclude from Equation (4.47) that

$$\det \gamma_\mu = \det(-\gamma_\mu) \tag{4.48}$$

But

$$\det(-\gamma_\mu) = (-1)^N \det \gamma_\mu \tag{4.49}$$

where N is the order of the γ matrix. The relation in Equation (4.49) can be verified using the fact that the determinant of a matrix of order N is a sum of terms where each term is the product of N factors. The negative of a matrix has the signs of all its elements inverted, multiplying each term in the sum by $(-1)^N$. We conclude from Equation (4.48) and Equation (4.49) that the order N of the γ matrices must be even. But $N = 2$ does not work because even though the three Pauli matrices satisfy the requirements discussed above, we need four linearly independent γ matrices and the fourth matrix that completes the set of four linearly independent matrices of second order is the unit matrix, and that matrix does not have zero trace.

Next we try γ matrices of fourth order. The four γ matrices are taken to be

$$\gamma_i = \begin{pmatrix} 0 & -i\sigma_i \\ i\sigma_i & 0 \end{pmatrix} \qquad \gamma_4 = \begin{pmatrix} I & 0 \\ 0 & -I \end{pmatrix} \tag{4.50}$$

where σ_i are the three Pauli matrices defined in Equation (4.19), I is the unit matrix of second order, and 0 the zero matrix of second order. The 4×4 γ matrices are seen to contain four 2×2 matrices. The four γ matrices are called 'block-diagonal'. When one multiplies two block-diagonal matrices, one can apply the usual rules of matrix multiplication using the blocks as matrix elements, ignoring that the blocks themselves are matrices. The many zero matrices involved make that multiplication even easier.

Because the Pauli matrices satisfy Equation (4.20), the γ_i $(i = 1, 2, 3)$ in Equation (4.50) automatically satisfy Equation (4.23). One can verify that $\{\gamma_4, \gamma_i\} = 0$ and $\gamma_4^2 = 1$ (here 1 is understood to be the 4×4 unit matrix) so the relations Equation (4.23) are satisfied by all γ_μ defined in Equation (4.50). One can also see that the γ matrices thus defined satisfy Equation (4.39), Equation (4.40), Equation (4.41) and Equation (4.46). The γ matrices defined in Equation (4.23) also satisfy the commutation relations

$$[\gamma_i, \gamma_j] = 2i\varepsilon_{ijk}\Sigma_k \tag{4.51}$$

with

$$\Sigma_k = \begin{pmatrix} \sigma_k & 0 \\ 0 & \sigma_k \end{pmatrix} \tag{4.52}$$

This relation can be proven by using

$$[\sigma_i, \sigma_j] = 2i\varepsilon_{ijk}\sigma_k \tag{4.53}$$

where the ε_{ijk} is the totally anti-symmetric Levi-Civita symbol. Adding Equation (4.23) and Equation (4.51) we obtain

$$\gamma_i\gamma_j = \delta_{ij} + i\varepsilon_{ijk}\Sigma_k \tag{4.54}$$

Because $\not{a}$ in Equation (4.26) and $\not{\partial}$ in Equation (4.27) are the sum of products of a matrix of fourth order and a scalar, $\not{a}$ and $\not{\partial}$ are themselves matrices of fourth order. It is understood that the second term in Equation (4.27) is multiplied by the unit matrix of fourth order, but we shall not usually show that unit matrix explicitly. Thus the expression between parentheses in Equation (4.27) is a matrix of fourth order and therefore ψ must be a 4×1 column vector

$$\psi = \begin{pmatrix} \psi_1 \\ \psi_2 \\ \psi_3 \\ \psi_4 \end{pmatrix} \tag{4.55}$$

We will return to the interpretation of these four components of ψ in the next section.

That all four γ matrices are Hermitian is one advantage of the 'East Coast metric' over the 'West Coast metric'. When doing math with the γ matrices one has to get used to the somewhat curious notation that stretches our concepts of a vector and a dot product beyond what we are used to.

4.2.2 Probability Density and Current

We are now in a position to verify that the gymnastics of the previous section leads to a positive definite probability density as promised in the beginning of that section. We go through the same procedure as followed for the Klein-Gordon equation in the previous section, and indeed as was used in the case of the Schrödinger equation.

We noted in the previous section that ψ is a 4×1 column vector. The complex conjugate of ψ is now replaced by the Hermitian conjugate $\psi^\dagger$

$$\psi^\dagger = \begin{pmatrix} \psi_1^* & \psi_2^* & \psi_3^* & \psi_4^* \end{pmatrix} \tag{4.56}$$

because taking the Hermitian conjugate of a matrix, whether square or not, entails transposing the matrix and replacing each element by its complex conjugate. Using Equation (4.55) and Equation (4.56) we find for the product $\psi^\dagger\psi$

$$\psi^\dagger\psi = \psi_1^*\psi_1 + \psi_2^*\psi_2 + \psi_3^*\psi_3 + \psi_4^*\psi_4 \tag{4.57}$$

Note that according to the rule for matrix multiplication, the multiplication of a 1×4 row matrix and a 4×1 column matrix results in a scalar quantity. It is seen from Equation (4.57) that $\psi^\dagger\psi$ is the sum of terms that are probability densities in the case of the Schrödinger equation and the Klein-Gordon equation, so it seems reasonable to interpret $\psi^\dagger\psi$ as a generalized probability density (probability per unit volume). It is seen to be positive definite, a requirement for such an interpretation in the absence of second quantization.

We now derive the form of the probability current and law of conservation of probability from the Dirac equation. The argument follows closely those in Section 4.1. We start with Equation (4.27) and write its spatial and time components separately

$$\gamma_i \frac{\partial \psi}{\partial x_i} + \gamma_4 \frac{\partial \psi}{\partial x_4} + m\psi = 0 \tag{4.58}$$

where, as before, repeated summation indices indicate a summation, from 1–3 for a Latin summation index and 1–4 for a Greek summation index. To isolate the term with the time derivative we multiply Equation (4.58) from the left by γ_4 to obtain with $\gamma_4^2 = 1$

$$\gamma_4 \gamma_i \frac{\partial \psi}{\partial x_i} + \frac{1}{i} \frac{\partial \psi}{\partial t} + m\gamma_4 \psi = 0 \tag{4.59}$$

and multiply it from the left by $\psi^\dagger$ to obtain

$$\psi^\dagger \gamma_4 \gamma_i \frac{\partial \psi}{\partial x_i} + \frac{1}{i} \psi^\dagger \frac{\partial \psi}{\partial t} + m\psi^\dagger \gamma_4 \psi = 0 \tag{4.60}$$

Next we take the complex conjugate of Equation (4.59) to get

$$\frac{\partial \psi^\dagger}{\partial x_i} \gamma_i^\dagger \gamma_4^\dagger - \frac{1}{i} \frac{\partial \psi^\dagger}{\partial t} + m\psi^\dagger \gamma_4^\dagger = 0 \tag{4.61}$$

and multiply it from the right by ψ to get

$$\frac{\partial \psi^\dagger}{\partial x_i} \gamma_i \gamma_4 \psi - \frac{1}{i} \frac{\partial \psi^\dagger}{\partial t} \psi + m\psi^\dagger \gamma_4 \psi = 0 \tag{4.62}$$

We have used that the Hermitian conjugate of a product of matrices is the product of their Hermitian conjugates but in reverse order, and that the γ matrices are Hermitian. We subtract Equation (4.62) from Equation (4.60) to get

$$\psi^\dagger \gamma_4 \gamma_i \frac{\partial \psi}{\partial x_i} - \frac{\partial \psi^\dagger}{\partial x_i} \gamma_i \gamma_4 \psi + \frac{1}{i} \left(\psi^\dagger \frac{\partial \psi}{\partial t} + \frac{\partial \psi^\dagger}{\partial t} \psi \right) = 0 \tag{4.63}$$

or

$$\frac{\partial}{\partial x_i} (\psi^\dagger \gamma_4 \gamma_i \psi) + \frac{1}{i} \frac{\partial}{\partial t} (\psi^\dagger \psi) = 0 \tag{4.64}$$

where we used that $\gamma_i \gamma_4 = -\gamma_4 \gamma_i$. We define the probability density ρ using the positive definite expression in Equation (4.57)

$$\rho = \psi^\dagger \psi \tag{4.65}$$

To obtain conservation of probability, see Equation (4.12), we define the probability current density as

$$j_k = i\psi^\dagger \gamma_4 \gamma_k \psi \tag{4.66}$$

where we replace the summation index i by k, leaving i to be used as the imaginary unit. Remember that one can use Gauss' theorem to show that Equation (4.12) and Equation (4.13) imply conservation of probability. The problem that Dirac set out to solve is now taken care of but at the expense of a wave function that is a column vector with four components. Before we provide an interpretation of these four components we rewrite the probability density and current density using $\overline{\psi}$ defined as

$$\overline{\psi} = \psi^\dagger \gamma_4 \tag{4.67}$$

Note that we multiply a 1×4 row vector by a 4×4 matrix. The resulting vector is also a 1×4 row vector. With this definition and using Equation (4.65) for the probability density ρ we get

$$\rho = \psi^\dagger \psi = \psi^\dagger \gamma_4 \gamma_4 \psi = \overline{\psi} \gamma_4 \psi \tag{4.68}$$

Similarly using Equation (4.66) for the probability current j_i we get

$$j_k = i\psi^\dagger \gamma_4 \gamma_k \psi = i\overline{\psi} \gamma_k \psi \tag{4.69}$$

The relations in Equation (4.68) and Equation (4.69) can be combined in a Lorenz vector $j_\mu = (\mathrm{j}, i\rho)$ with

$$j_\mu = i\overline{\psi} \gamma_\mu \psi \tag{4.70}$$

Note the factor i. The conservation of probability is now expressed as

$$\partial_\mu j_\mu = 0 \tag{4.71}$$

as usual, compare with Equation (4.13). The electromagnetic current density is naturally defined as charge times probability current density or

$$j_\mu^{\text{em}} = iq\overline{\psi} \gamma_\mu \psi \tag{4.72}$$

where q is the charge of the particle described by ψ. We shall have use for it later. We note that j_μ in Equation (4.70) is a Lorenz four-vector and that Equation (4.71) is Lorenz covariant as required if the formalism is to be covariant.

We can cast the Dirac Equation (4.27) for ψ in an equation valid for $\overline{\psi}$ by taking the Hermitian conjugate of Equation (4.27) to obtain

$$\psi^\dagger (\not{\partial}^\dagger + m) = 0 \tag{4.73}$$

When we multiply this equation from the right by γ_4 we get

$$\psi^\dagger(\not{\partial}^\dagger\gamma_4 + m\gamma_4) = 0 \tag{4.74}$$

We would like to commute γ_4 to the left so that it is adjacent to $\psi^\dagger$ and that we get $\psi^\dagger\gamma_4 = \overline{\psi}$. Using the fact that $\not{\partial}^\dagger = (\gamma_\mu\partial_\mu)^\dagger = \gamma_\mu^\dagger p_\mu^* = \gamma_\mu p_\mu^*$ and that $p_i^* = p_i$ and $p_4^* = (iE)^* = -iE = -p_4$ we can write Equation (4.74) as

$$\psi^\dagger(\gamma_i p_i \gamma_4 - \gamma_4 p_4 \gamma_4 + m\gamma_4) = 0 \tag{4.75}$$

or

$$\psi^\dagger\gamma_4(-\gamma_i p_i - \gamma_4 p_4 + m) = 0 \tag{4.76}$$

so we get

$$\overline{\psi}(\not{\partial} - m) = 0 \tag{4.77}$$

The operator $\not{\partial}$ is understood to operate on the quantity on its left. The equation for $\overline{\psi}$ differs from the equation for ψ by a minus sign in front of the mass term.

4.3 SOLUTIONS OF THE DIRAC EQUATION, ANTI-PARTICLES

We will now solve the Dirac equation and interpret the solutions. We will find that the solutions describe spin $\frac{1}{2}$ particles and spin $\frac{1}{2}$ anti-particles each with a spin up and a spin down component, corresponding to the four components of the wave function in Equation (4.55).

4.3.1 Solutions of the Dirac Equation

The Dirac equation does not seem to include interactions, so we will look for solutions that correspond to free particles propagating in space. Thus we try plane wave solutions of the form

$$\psi(x) = u\, e^{ipx} \tag{4.78}$$

where p and x are the momentum and position four-vectors and u may depend upon p but not upon x. Substitution of Equation (4.78) in the Dirac Equation (4.27) gives

$$(i\not{p} + m)\, u(p) = 0 \tag{4.79}$$

In view of Equation (4.55), u is given by a 4×1 column vector

$$u = \begin{pmatrix} u_{A1} \\ u_{A2} \\ u_{B1} \\ u_{B2} \end{pmatrix} \tag{4.80}$$

The column vector u is called a spinor. The names of the indices in Equation (4.80) are chosen as shown because the fourth order γ matrices have a block structure where the blocks are second-order matrices, see Equation (4.50). Therefore we also write Equation (4.80) in a corresponding block structure as

$$u = \begin{pmatrix} u_A \\ u_B \end{pmatrix} \tag{4.81}$$

where u_A and u_B are given by

$$u_A = \begin{pmatrix} u_{A1} \\ u_{A2} \end{pmatrix} \qquad u_B = \begin{pmatrix} u_{B1} \\ u_{B2} \end{pmatrix} \tag{4.82}$$

We substitute Equation (4.82) in the Dirac equation written in the form Equation (4.59). For the substitution we evaluate the derivatives as follows

$$\frac{\partial}{\partial x_i}\left(u\, \mathrm{e}^{ipx}\right) = ip_i u\, \mathrm{e}^{ipx} \qquad \frac{\partial}{\partial x_4}\left(u\, \mathrm{e}^{ipx}\right) = \frac{1}{i}(-iE)\, u\, \mathrm{e}^{ipx} \tag{4.83}$$

Substitution of the relations Equation (4.83) into Equation (4.59) and using Equation (4.50) gives

$$\begin{pmatrix} 0 & -i\sigma_i \\ i\sigma_i & 0 \end{pmatrix} ip_i u\, \mathrm{e}^{ipx} + \begin{pmatrix} 1 & 0 \\ 0 & -1 \end{pmatrix} (-E) u\, \mathrm{e}^{ipx} + \begin{pmatrix} 1 & 0 \\ 0 & 1 \end{pmatrix} mu\, \mathrm{e}^{ipx} = 0 \tag{4.84}$$

or

$$\begin{pmatrix} m - E & \boldsymbol{\sigma} \cdot \mathbf{p} \\ -\boldsymbol{\sigma} \cdot \mathbf{p} & m + E \end{pmatrix} \begin{pmatrix} u_A \\ u_B \end{pmatrix} = 0 \tag{4.85}$$

The elements on the diagonal are understood to be multiplied by the unit matrix of second order. This calculation shows how to do algebra with blocked matrices. It is crucial that the reader verify that the matrices of fourth order can indeed be multiplied as if they are matrices of second order, that is, one does not need to make use of the fact that the blocks are themselves matrices of second order when executing the multiplication.

We see that Equation (4.85) represents a set of four linear homogeneous equations. We will do the algebra as if Equation (4.85) represents two linear homogeneous equations with u_A and u_B the two unknowns. Linear homogeneous equations have a non-trivial solution if the determinant of the coefficients is 0. Thus we require

$$(m - E)(m + E) + (\boldsymbol{\sigma} \cdot \mathbf{p})^2 = 0 \tag{4.86}$$

The quantity $(\boldsymbol{\sigma} \cdot \mathbf{p})^2$ can be evaluated using

$$\sigma_i \sigma_j = \delta_{ij} + i\varepsilon_{ijk}\sigma_k \tag{4.87}$$

compare with Equation (4.54). So

$$(\boldsymbol{\sigma} \cdot \mathbf{p})^2 = \sigma_i p_i \sigma_j p_j = (\delta_{ij} + \varepsilon_{ijk}\sigma_k)\, p_i p_j = \mathbf{p}^2 \tag{4.88}$$

where we have used that $\varepsilon_{ijk} p_i p_j = 0$. This follows from the fact that ε_{ijk} is anti-symmetric and $p_i p_j$ is symmetric. In general $A_{ij} S_{ij} = 0$ if A_{ij} is anti-symmetric and S_{ij} is symmetric. This can be seen as folllows

$$A_{ij} S_{ij} = A_{ij} S_{ji} = -A_{ji} S_{ji} = -A_{ij} S_{ij} \tag{4.89}$$

where in the first equality we used that S is symmetric, in the second equality that A is antisymmetric and in the third equality we simply replaced the dummy summation variables i and j by j and i. The property holds as well if there are more indices. Using Equation (4.88) in Equation (4.86) we find

$$m^2 - E^2 + \mathbf{p}^2 = 0 \tag{4.90}$$

This is the Einstein relation and it is of course satisfied by $E$and $\mathbf{p}$, guaranteeing that non-trivial solutions for Equation (4.85) exist. The set of linear homogeneous Equations (4.85) are degenerate when the condition Equation (4.90) is satisfied. This means that the best we can do is to get one solution in terms of the other.

Consider for simplicity a particle at rest. Then $E = \pm m$ and Equation (4.85) reduces to

$$(m - E)\, u_A = 0 \tag{4.91}$$

$$(m + E)\, u_B = 0 \tag{4.92}$$

Each of these two equations represents two identical equations because they are understood to contain the two-dimensional unit matrix. If $E = +m > 0$ we see from Equation (4.91) that u_A is undetermined and from Equation (4.92) that $u_B = 0$. We choose $u_A = 1$. This is allowed because if solutions of linear and homogeneous equations are multiplied by an arbitrary constant, they remain solutions. Because u_A and u_B are 2×1 column vectors this means that

$$u_A = \begin{pmatrix} 1 \\ 0 \end{pmatrix} \quad \text{or} \quad u_A = \begin{pmatrix} 0 \\ 1 \end{pmatrix} \tag{4.93}$$

and

$$u_B = \begin{pmatrix} 0 \\ 0 \end{pmatrix} \tag{4.94}$$

When $E = -m < 0$ we see from Equation (4.91) that $u_A = 0$ and from Equation (4.92) that u_B is undetermined. We choose $u_B = 1$ which means

$$u_B = \begin{pmatrix} 1 \\ 0 \end{pmatrix} \quad \text{or} \quad u_B = \begin{pmatrix} 0 \\ 1 \end{pmatrix} \tag{4.95}$$

and

$$u_A = \begin{pmatrix} 0 \\ 0 \end{pmatrix} \tag{4.96}$$

When $\mathbf{p} \neq 0$ we write Equation (4.85) in full as

$$(m - E)\, u_A + \boldsymbol{\sigma} \cdot \mathbf{p}\, u_B = 0 \tag{4.97}$$

$$-\boldsymbol{\sigma} \cdot \mathbf{p}\, u_A + (m + E)\, u_B = 0 \tag{4.98}$$

Again, each of these two equations represents two equations. The terms with $(m - E)u_A$ and $(m + E)u_b$ are understood to contain the two-dimensional unit matrix. It is illustrative to write them out in full to verify that (symbolic) division by $(m - E)$ and $(m + E)$ respectively is allowed (we are *not* dividing by the unit matrix). To decide whether to express u_A in terms of u_B or u_B in terms of u_A we consider that $\boldsymbol{\sigma} \cdot \mathbf{p}$ must appear in the numerator because one cannot divide by a matrix (and we do not want to introduce the inverse of $\boldsymbol{\sigma}$ needlessly). When $E \geq 0$, putting $E - m$ in the denominator can lead to a division by 0 if $\mathbf{p} = 0$. Thus we use Equation (4.98) to solve for u_B in terms of u_A to find

$$u_B = \frac{\boldsymbol{\sigma} \cdot \mathbf{p}}{m + E} u_A \ \text{with} \ u_A = \begin{pmatrix} 1 \\ 0 \end{pmatrix} \ \text{or} \ u_A = \begin{pmatrix} 0 \\ 1 \end{pmatrix} \qquad E > 0 \tag{4.99}$$

Similarly, when $E \leq 0$, putting $E + m$ in the denominator can lead to a division by 0 if $\mathbf{p} = 0$. Thus we use Equation (4.97) to solve for u_A in terms of u_B to find

$$u_A = -\frac{\boldsymbol{\sigma} \cdot \mathbf{p}}{m - E} u_B \ \text{with} \ u_B = \begin{pmatrix} 1 \\ 0 \end{pmatrix} \ \text{or} \ u_B = \begin{pmatrix} 0 \\ 1 \end{pmatrix} \qquad E < 0 \tag{4.100}$$

Note that for $\mathbf{p} = 0$, Equation (4.99) and Equation (4.100) are identical to Equation (4.93), Equation (4.94) and Equation (4.95), Equation (4.96) respectively.

We summarize the four solutions found as

$$u_1(E,\mathbf{p}) = N\begin{pmatrix} \begin{pmatrix} 1 \\ 0 \end{pmatrix} \\ \frac{\sigma\cdot\mathbf{p}}{m+E}\begin{pmatrix} 1 \\ 0 \end{pmatrix} \end{pmatrix} \qquad u_2(E,\mathbf{p}) = N\begin{pmatrix} \begin{pmatrix} 0 \\ 1 \end{pmatrix} \\ \frac{\sigma\cdot\mathbf{p}}{m+E}\begin{pmatrix} 0 \\ 1 \end{pmatrix} \end{pmatrix} \qquad E > 0 \tag{4.101}$$

$$u_3(E,\mathbf{p}) = N\begin{pmatrix} -\frac{\sigma\cdot\mathbf{p}}{m+E}\begin{pmatrix} 1 \\ 0 \end{pmatrix} \\ \begin{pmatrix} 1 \\ 0 \end{pmatrix} \end{pmatrix} \qquad u_4(E,\mathbf{p}) = N\begin{pmatrix} -\frac{\sigma\cdot\mathbf{p}}{m+E}\begin{pmatrix} 0 \\ 1 \end{pmatrix} \\ \begin{pmatrix} 0 \\ 1 \end{pmatrix} \end{pmatrix} \qquad E < 0 \tag{4.102}$$

We have introduced a (normalization) constant N whose value is yet to be chosen. We call u_1 and u_2 the positive energy spinors and u_3 and u_4 the negative energy spinors. The expressions for the four spinors show a clear pattern but the pattern is broken by the negative signs appearing in u_3 and in u_4 in Equation (4.102) as compared to u_1 and u_2 in Equation (4.101). This will be alleviated in the next subsection.

4.3.2 Anti-Particles

The negative energy solutions are due to the fact that, for given $\mathbf{p}$, the Einstein equation

$$E^2 = \mathbf{p}^2 + m^2 \tag{4.103}$$

has two solutions for the energy, one positive and one negative, and this cannot be avoided. Dirac realized that the physical world would not be stable: all particles would fall into lower negative energy levels by spontaneous emission of a photon, and this process would go on without limit because there are an infinite number of levels with negative energy available. We shall show in the next section that the Dirac equation describes spin $\frac{1}{2}$ particles. Dirac postulated that normally *all* negative energy states are completely filled with these particles, obeying the Pauli exclusion principle. We ignore for the moment the sticky question about what the vacuum looks like when filled with an infinite number of particles that, when charged, repel each other. The region $-m < E < +m$ is excluded because of Equation (4.103). This 'solves' the stability problem of the physical world because the Pauli exclusion principle does not permit particles occupying a positive energy level to fall into an already occupied negative energy level. But it would be possible for a photon to collide with one of the particles in a negative energy level and bring that particle to a positive energy level, provided that level is not occupied with a particle with the same quantum numbers. It is clear that that photon must have a minimum energy of $2m$ for this process to be kinematically allowed. If an electron is brought from a negative energy level to a positive energy level, a 'hole' remains in the otherwise filled negative energy levels. This configuration with the hole, when compared with the prior situation, appears to have a particle whose quantum numbers are the opposite from the particle that was removed. For example, if the removed particle had an energy E, the energy of the hole appears to be $-E$. Likewise, if the removed particle was charged, the apparent charge of the hole will have the opposite sign of that charge. The apparent z-component of the angular momentum of a hole will be the opposite from the z-component of the angular momentum of the removed particle. When a particle with negative energy moves into a hole, the hole appears to have moved in the opposite direction. Therefore, the velocity and the momentum of the hole are opposite of those of the moving particle. We call particles with such opposite quantum numbers anti-particles.

The Dirac equation leads to the prediction of anti-particles. This prediction was borne out within a few years with the discovery of the positron and later the identification of an anti-particle for every known particle (some particles are their own anti-particle). We are led to interpret particles at negative energy as anti-particles whose momenta and energy are the opposite of particles, that is, the four-vector p is replaced by $-p$. We make the replacements $\mathbf{p} \to -\mathbf{p}$ and $E \to -|E|$ for the negative energy solutions u_3 and u_4 in Equation (4.102) to get

$$u_3(-|E|, -\mathbf{p}) = N \begin{pmatrix} \frac{\sigma\cdot\mathbf{p}}{m+|E|}\begin{pmatrix}1\\0\end{pmatrix} \\ \begin{pmatrix}1\\0\end{pmatrix} \end{pmatrix} \qquad u_4(-|E|, -\mathbf{p}) = N \begin{pmatrix} \frac{\sigma\cdot\mathbf{p}}{m+|E|}\begin{pmatrix}0\\1\end{pmatrix} \\ \begin{pmatrix}0\\1\end{pmatrix} \end{pmatrix} \tag{4.104}$$

We write $|E|$ to remind ourselves that we must use a positive value for the energy when evaluating u_3 and u_4. The expressions for the four spinors now show a clear pattern which aids in memorizing them. The exception is the appearance if $|E|$ instead of E (of course we could write $|E|$ in the expressions Equation (4.101) for u_1 and u_2).

It is customary to introduce new spinors v_1 and v_2 instead of the spinors u_3 and u_4 defined in Equation (4.104). They are defined as

$$v_1(|E|, \mathbf{p}) = -u_4(-|E|, -\mathbf{p}) \tag{4.105}$$

$$v_2(|E|, \mathbf{p}) = u_3(-|E|, -\mathbf{p}) \tag{4.106}$$

The minus sign in front of u_4 in Equation (4.105) is conventional and is allowed, of course, because the spinors are solutions of a set of linear and homogeneous equations. It affects the scalar product of the spinors with each other, as will be seen in Problem 2 at the end of this Chapter. The connection between v_1 and u_4 on the one hand and v_2 and u_3 on the other has to do with the labeling of the intrinsic angular momentum (spin) discussed in the next section. From here on it is assumed that E is a positive quantity and that $\mathbf{p}$ is the momentum of either particle in the case of u or the anti-particle in the case of v, and we will write E instead of $|E|$ and $\mathbf{p}$ instead of $-\mathbf{p}$ for anti-particles. Thus the spinors v_1 and v_2 take the form

$$v_1(E,\mathbf{p}) = -N\begin{pmatrix} \frac{\sigma\cdot\mathbf{p}}{m+E}\begin{pmatrix}0\\1\end{pmatrix} \\ \begin{pmatrix}0\\1\end{pmatrix}\end{pmatrix} \qquad v_2(E,\mathbf{p}) = N\begin{pmatrix} \frac{\sigma\cdot\mathbf{p}}{m+E}\begin{pmatrix}1\\0\end{pmatrix} \\ \begin{pmatrix}1\\0\end{pmatrix}\end{pmatrix} \tag{4.107}$$

The near symmetry between the u and v spinors is apparent.

With the substitution $p \to -p$ in u_3 and u_4 to obtain v_2 and v_1 respectively, we see from Equation (4.79) that v_1 and v_2 satisfy

$$(i\not{p} - m)\,v(p) = 0 \tag{4.108}$$

while u_1 and u_2 continue to satisfy Equation (4.79)

$$(i\not{p} + m)\,u(p) = 0 \tag{4.109}$$

Note the sign difference between Equation (4.108) and Equation (4.109).

We can derive equations for $\overline{u}(p)$ and $\overline{v}(p)$ from Equation (4.109) and Equation (4.108) following the procedure we used to derive Equation (4.77) for $\overline{\psi}$ from Equation (4.27) for ψ. We take the Hermitian conjugate of Equation (4.109) and Equation (4.108) and multiply the resulting equations from the right by γ_4. We commute γ_4 to its left until it is adjacent to $u^\dagger$ or $v^\dagger$. We obtain

$$\overline{u}(p)\,(i\gamma p + m) = 0 \tag{4.110}$$

$$\overline{v}(p)\,(i\gamma p - m) = 0 \tag{4.111}$$

The equations for u and $\overline{u}$ are the same, as are the equations for v and $\overline{v}$, except for the fact that the u and v are trailing and the $\overline{u}$ and $\overline{v}$ are leading the dot products γp.

We leave the normalization of the spinors for Problem 2 at the end of this chapter.Now that we have understood a doubling of the number of components of the wave function, the remaining question is why the wavefunction has four instead of two components.

4.4 SPIN, NON-RELATIVISTIC LIMIT AND MAGNETIC MOMENT

4.4.1 Orbital Angular Momentum

In the Dirac theory orbital angular momentum is not conserved. This shocking fact can be verified by calculating the commutator of the orbital angular momentum $\mathbf{L} = \mathbf{x} \times \mathbf{p}$ or $L_i = \varepsilon_{ijk} x_j p_k$ and the Hamiltonian $H = \alpha_i p_i + \beta m$ of Equation (4.32)

$$\begin{aligned}
[L_i, H] &= \varepsilon_{ijk} x_j p_k (\alpha_l p_l + \beta m) - (\alpha_l p_l + \beta m) \varepsilon_{ijk} x_j p_k \\
&= \varepsilon_{ijk} \alpha_l (x_j p_k p_l - p_l x_j p_k) \\
&= \varepsilon_{ijk} \alpha_l \left(x_j p_k p_l - x_j p_l p_k - \frac{1}{i} \delta_{lj} p_k \right) \\
&= i \varepsilon_{ijk} \alpha_l \delta_{lj} p_k \\
&= i \varepsilon_{ijk} \alpha_j p_k \\
&= i (\boldsymbol{\alpha} \times \mathbf{p})_i
\end{aligned} \tag{4.112}$$

We have used the commutator relation $[p_l, x_j] = \delta_{ij/i}$ in the second line. The commutator relation can also be written as

$$[\mathbf{L}, H] = i \boldsymbol{\alpha} \times \mathbf{p} \tag{4.113}$$

In general the commutator is not zero and therefore in general $\mathbf{L}$ and H do not share eigenkets and the eigenvalues of $\mathbf{L}$ are not good quantum numbers (see Chapter 3, Section 3.1). This latter conclusion is borne out by the fact that

$$\frac{d\mathbf{L}}{dt} = i [H, \mathbf{L}] = \boldsymbol{\alpha} \times \mathbf{p} \tag{4.114}$$

so the expectation value of $\mathbf{L}$ is not conserved. Thus we conclude that orbital angular momentum is not conserved in the Dirac theory. This is worrisome to say the least. In the language of Section 3.2, rotations of the coordinate system generated by $\mathbf{L}$ are not symmetry operations!

Is momentum conserved? We have

$$\frac{d\mathbf{p}}{dt} = i[H, \mathbf{p}] = i [\boldsymbol{\alpha} \cdot \mathbf{p} + \beta m, \mathbf{p}] = 0 \tag{4.115}$$

so momentum is indeed conserved and the momentum operator shares eigenkets with the Hamiltonian and translations of the coordinate system are symmetry operations.

Before addressing the orbital angular momentum issue, we note an interesting exception to the above statement with regard to orbital angular momentum. Because $\boldsymbol{\alpha} \times \mathbf{p} \cdot \mathbf{p} = 0$ we are led to consider the quantity $\mathbf{L} \cdot \mathbf{p}$. We find using Equation (4.113)

$$\frac{d\mathbf{L} \cdot \mathbf{p}}{dt} = i[H, \mathbf{L} \cdot \mathbf{p}] = \boldsymbol{\alpha} \times \mathbf{p} \cdot \mathbf{p} = 0 \tag{4.116}$$

So using the language of Section 3.2, rotations of the coordinate system generated by $\mathbf{L}$ around an axis in the direction of $\mathbf{p}$ are symmetry operations, and the component of $\mathbf{L}$ along $\mathbf{p}$ is conserved. This is intricately connected to the helicity of the particle, see below.

4.4.2 Spin and Total Angular Momentum

To restore the conservation of angular momentum we seek to add a term to the orbital angular momentum $\mathbf{L}$ so that the total angular momentum, called $\mathbf{J}$, is conserved. Of course this additional term will correspond to the intrinsic angular momentum or spin of the particle. At the time Dirac wrote down his equation, the existence of spin was already suggested by experiment. In analogy with the spin postulated in non-relativistic quantum theory, we investigate an additional term of the form of Equation (4.52)

$$\boldsymbol{\Sigma} = \begin{pmatrix} \boldsymbol{\sigma} & 0 \\ 0 & \boldsymbol{\sigma} \end{pmatrix} \tag{4.117}$$

and anticipate that the extra term that corresponds to spin will be $\boldsymbol{\Sigma}/2$, just as in non-relativistic Quantum Theory $\mathbf{S} = \boldsymbol{\sigma}/2$. We calculate the commutator of $\boldsymbol{\Sigma}$ and the Hamiltonian $H = \boldsymbol{\alpha} \cdot \mathbf{p} + \beta m$

$$\begin{aligned}
[\Sigma_k, H] &= [\Sigma_k, (\alpha_l p_l + \beta m)] \\
&= [\Sigma_k, \alpha_l p_l] \\
&= \left[\begin{pmatrix} \sigma_k & 0 \\ 0 & \sigma_k \end{pmatrix}, \begin{pmatrix} 0 & \sigma_l \\ \sigma_l & 0 \end{pmatrix} p_l \right] \\
&= \begin{pmatrix} 0 & \sigma_k \sigma_l \\ \sigma_k \sigma_l & 0 \end{pmatrix} p_l - \begin{pmatrix} 0 & \sigma_l \sigma_k \\ \sigma_l \sigma_k & 0 \end{pmatrix} p_l \\
&= \begin{pmatrix} 0 & [\sigma_k, \sigma_l] \\ [\sigma_k, \sigma_l] & 0 \end{pmatrix} p_l \\
&= \begin{pmatrix} 0 & 2i\varepsilon_{klm}\sigma_m \\ 2i\varepsilon_{klm}\sigma_m & 0 \end{pmatrix} p_l \\
&= 2i\varepsilon_{klm} p_l \begin{pmatrix} 0 & \sigma_m \\ \sigma_m & 0 \end{pmatrix}
\end{aligned}$$

$$= 2i(\mathbf{p} \times \boldsymbol{\alpha})_k$$
$$= -2i(\boldsymbol{\alpha} \times \mathbf{p})_k \tag{4.118}$$

We have used that $[\Sigma_k, \beta] = 0$ because a diagonal matrix like β commutes with any matrix such as Σ_k and we have used Equation (4.53) for the commutator $[\sigma_k, \sigma_l]$. That α_l is given by

$$\alpha_l = \begin{pmatrix} 0 & \sigma_l \\ \sigma_l & 0 \end{pmatrix} \tag{4.119}$$

can be seen from the definition in Equation (4.33) of α_l and the expressions in Equation (4.50) for γ_l and γ_4. The relation in Equation (4.118) can also be written as

$$[\boldsymbol{\Sigma}, H] = -2i\boldsymbol{\alpha} \times \mathbf{p} \tag{4.120}$$

When we compare Equation (4.120) and Equation (4.114) we see that their right-hand sides cancel when we evaluate $[(\mathbf{L} + \frac{1}{2}\boldsymbol{\Sigma}), H]$ so with the definitions

$$\mathbf{J} = \mathbf{L} + \mathbf{S} \quad \text{and} \quad \mathbf{S} = \tfrac{1}{2}\boldsymbol{\Sigma} \tag{4.121}$$

we get

$$[\mathbf{J}, H] = 0 \tag{4.122}$$

We call $\mathbf{J}$ defined in Equation (4.121) the total angular momentum. Because $\mathbf{J}$ commutes with the Hamiltonian, $\mathbf{J}$ is a symmetry operator and we have restored the conservation of angular momentum. We will show later that $\frac{1}{2}\boldsymbol{\Sigma}$ indeed corresponds to a spin $\frac{1}{2}$ particle as expected from Equation (4.117).

The spin operator $\mathbf{S}$ is not a symmetry operator because it does not commute with the Hamiltonian and therefore its eigenvalues are not conserved. Neither are those of the orbital angular momentum $\mathbf{L}$, but the eigenvalues of the total angular momentum $\mathbf{J} = \mathbf{L} + \mathbf{S}$ *are* conserved. The spectroscopic notations used in atomic and nuclear physics are in principle incorrect!

We note an interesting exception to the above statement with regard to the spin. Because $\boldsymbol{\alpha} \times \mathbf{p} \cdot \mathbf{p} = 0$ we are led to consider the quantity $\mathbf{S} \cdot \mathbf{p}$. We find using Equation (4.120)

$$\frac{d\mathbf{S} \cdot \mathbf{p}}{dt} = i[H, \mathbf{S} \cdot \mathbf{p}] = -\boldsymbol{\alpha} \times \mathbf{p} \cdot \mathbf{p} = 0 \tag{4.123}$$

So using the language of Section 3.2, rotations of the coordinate system generated by $\mathbf{S}$ around an axis in the direction of $\mathbf{p}$ are symmetry operations, and the component of $\mathbf{S}$ along $\mathbf{p}$ is conserved. Compare with the discussion of $\mathbf{L} \cdot \mathbf{p}$ and Equation (4.116).

4.4.3 Helicity

This leads us to introduce the helicity operator for particles with either positive or negative energy as

$$h = \frac{\mathbf{\Sigma} \cdot \mathbf{p}}{|\mathbf{p}|} \tag{4.124}$$

which corresponds to the projection of the spin operator $\mathbf{S} = \mathbf{\Sigma}/2$ on the direction of $\mathbf{p}$. Evidently h is a 4×4 matrix. It follows from Equation (4.123) that the helicity operator commutes with the Hamiltonian H. Therefore it shares its eigenkets with H and its eigenvalues are conserved. Using Equation (4.117) for $\mathbf{\Sigma}$ we find that we can also write

$$h = \frac{1}{|\mathbf{p}|} \begin{pmatrix} \boldsymbol{\sigma} \cdot \mathbf{p} & 0 \\ 0 & \boldsymbol{\sigma} \cdot \mathbf{p} \end{pmatrix} \tag{4.125}$$

To find the eigenvalues of the helicity operator we calculate h^2

$$h^2 = \frac{1}{\mathbf{p}^2} \begin{pmatrix} (\boldsymbol{\sigma} \cdot \mathbf{p})^2 & 0 \\ 0 & (\boldsymbol{\sigma} \cdot \mathbf{p})^2 \end{pmatrix} = \frac{1}{\mathbf{p}^2} \begin{pmatrix} \mathbf{p}^2 & 0 \\ 0 & \mathbf{p}^2 \end{pmatrix} = \begin{pmatrix} 1 & 0 \\ 0 & 1 \end{pmatrix} \tag{4.126}$$

where we used Equation (4.88) for $(\boldsymbol{\sigma} \cdot \mathbf{p})^2$. Thus the eigenvalues of the helicity operator are $h = \pm 1$. This implies that the spin of the particle is $\frac{1}{2}$ because only that value has exactly two values along an axis of quantization (here the direction of $\mathbf{p}$). We call these spin 'up' ($h = +1$) and spin 'down' ($h = -1$).

Which of the four spinors u_1, u_2, v_1 and v_2 have spin up and spin down can be found as follows. Assume a particle has momentum $\mathbf{p}$ and choose the z-axis along the direction of $\mathbf{p}$. Then $\boldsymbol{\sigma} \cdot \mathbf{p} = \sigma_3 p_3$, $|\mathbf{p}| = p_3$ and the helicity operator Equation (4.125) simplifies to

$$h = \frac{1}{p_3} \begin{pmatrix} \sigma_3 p_3 & 0 \\ 0 & \sigma_3 p_3 \end{pmatrix} = \begin{pmatrix} \sigma_3 & 0 \\ 0 & \sigma_3 \end{pmatrix} \tag{4.127}$$

We then find

$$h\, u_1 = \begin{pmatrix} 1 & & & \\ & -1 & & \\ & & 1 & \\ & & & -1 \end{pmatrix} \begin{pmatrix} \begin{pmatrix} 1 \\ 0 \end{pmatrix} \\ \frac{\boldsymbol{\sigma} \cdot \mathbf{p}}{m+E} \begin{pmatrix} 1 \\ 0 \end{pmatrix} \end{pmatrix} = \begin{pmatrix} \begin{pmatrix} 1 \\ 0 \end{pmatrix} \\ \frac{\boldsymbol{\sigma} \cdot \mathbf{p}}{m+E} \begin{pmatrix} 1 \\ 0 \end{pmatrix} \end{pmatrix} = +u_1 \tag{4.128}$$

$$h\, u_2 = \begin{pmatrix} 1 & & & \\ & -1 & & \\ & & 1 & \\ & & & -1 \end{pmatrix} \begin{pmatrix} \begin{pmatrix} 0 \\ 1 \end{pmatrix} \\ \frac{\boldsymbol{\sigma} \cdot \mathbf{p}}{m+E} \begin{pmatrix} 0 \\ 1 \end{pmatrix} \end{pmatrix} = \begin{pmatrix} \begin{pmatrix} 0 \\ -1 \end{pmatrix} \\ \frac{\boldsymbol{\sigma} \cdot \mathbf{p}}{m+E} \begin{pmatrix} 0 \\ -1 \end{pmatrix} \end{pmatrix} = -u_2 \tag{4.129}$$

An anti-particle with momentum $\mathbf{p}$ corresponds to a particle with negative energy and momentum $-\mathbf{p}$. Then $\boldsymbol{\sigma} \cdot \mathbf{p} = \sigma_3(-p_3)$, $|\mathbf{p}| = p_3$ and the helicity operator Equation (4.125) simplifies to

$$h = \frac{1}{p_3}\begin{pmatrix} \sigma_3(-p_3) & 0 \\ 0 & \sigma_3(-p_3) \end{pmatrix} = \begin{pmatrix} -\sigma_3 & 0 \\ 0 & -\sigma_3 \end{pmatrix} \tag{4.130}$$

We then find

$$h\,v_1 = \begin{pmatrix} -1 & & & \\ & 1 & & \\ & & -1 & \\ & & & 1 \end{pmatrix}(-1)\begin{pmatrix} \frac{\sigma \cdot p}{m+E}\begin{pmatrix} 0 \\ 1 \end{pmatrix} \\ \begin{pmatrix} 0 \\ 1 \end{pmatrix} \end{pmatrix} = -\begin{pmatrix} \frac{\sigma \cdot p}{m+E}\begin{pmatrix} 0 \\ 1 \end{pmatrix} \\ \begin{pmatrix} 0 \\ 1 \end{pmatrix} \end{pmatrix} = +v_1 \tag{4.131}$$

$$h\,v_2 = \begin{pmatrix} -1 & & & \\ & 1 & & \\ & & -1 & \\ & & & 1 \end{pmatrix}\begin{pmatrix} \frac{\sigma \cdot p}{m+E}\begin{pmatrix} 1 \\ 0 \end{pmatrix} \\ \begin{pmatrix} 1 \\ 0 \end{pmatrix} \end{pmatrix} = \begin{pmatrix} \frac{\sigma \cdot p}{m+E}\begin{pmatrix} -1 \\ 0 \end{pmatrix} \\ \begin{pmatrix} -1 \\ 0 \end{pmatrix} \end{pmatrix} = -v_2 \tag{4.132}$$

The minus sign between the helicity matrix and the spinor in Equation (4.131) is from the definition of v_1, see Equation (4.105) and Equation (4.107). The helicity assignments of v_1 and v_2 can be derived in a different manner using the relation between them and u_3 and u_4, see Equation (4.105) and Equation (4.106). It is easy to show that $hu_3(E, \mathbf{p}) = u_3(E, \mathbf{p})$ with $E < 0$. Making the replacements $E \to -E$ and $\mathbf{p} \to -\mathbf{p}$ we obtain $-hu_3(-E, -\mathbf{p}) = u_3(-E, -\mathbf{p})$ where we used the fact that h changes sign when $\mathbf{p} \to -\mathbf{p}$. But $u_3(-E, -\mathbf{p}) = v_2(E, \mathbf{p})$ so we get $-hv_2(E, \mathbf{p}) = v_2(E, -\mathbf{p})$ showing that v_2 has in negative helicity. A similar argument confirms that v_1 has positive helicity. It is obvious that the minus sign in the definition of v_1 in Equation (4.105) has no effect on the helicity of v_1. Thus we see that u_1 and v_1 (and u_3) have positive helicity and represent right-handed objects, and that u_2 and v_2 (and u_4) have negative helicity and represent left-handed objects. These results for the helicity assignment show why we chose to connect v_1 to u_4 (and thus v_2 to u_3) in the definitions Equation (4.105) and Equation (4.106). There are several conventions in the literature for defining v_1 and v_2 so care must be exercised when consulting the literature.

We conclude that the four components of the wavefunction ψ of Equation (4.80) correspond to spin up and spin down of a particle and spin up and spin down of an anti-particle. Thus the Dirac equation describes spin $\frac{1}{2}$ particles and anti-particles.

Initially Dirac postulated that the particles were electrons and that the anti-particles were protons. But Weyl showed that the anti-particles should have the same mass as the particles and Oppenheimer showed that with

Dirac's assignment, electrons and protons should annihilate each other in about 10^{-10} sec. Not a very healthy situation for the stability of atoms and nuclei. In 1932 Anderson discovered the positron, the anti-particle of the electron, using a cloud chamber exposed to gamma rays generated by cosmic rays. It took until 1955 before the anti-proton was discovered. It is now known that all particles either have an anti-particle or they are their own anti-particle. For the latter to be the case, the particle has to be neutral of course. An example of the latter is the photon.

4.4.4 Non-Relativistic Limit

In this section we take the Dirac equation to its non-relativistic limit and make contact with the Schrödinger equation. As a bonus we get an expression for the magnetic moment of a Dirac particle. The magnetic moment interacts with a magnetic field so we want to introduce the electromagnetic interaction in the Dirac equation. We do that with the subsitution in Equation (1.4)corresponding to the minimal substitution

$$p \to p - eA^{\mu} \tag{4.133}$$

with e the charge of the Dirac particle (often but not always an electron with $e < 0$). This can be recast as

$$\frac{1}{i}\frac{\partial}{\partial x_{\mu}} \to \frac{1}{i}\frac{\partial}{\partial x_{\mu}} - eA_{\mu} \tag{4.134}$$

or

$$\frac{1}{i}\gamma_{\mu}\frac{\partial}{\partial x_{\mu}} \to \frac{1}{i}\gamma_{\mu}\frac{\partial}{\partial x_{\mu}} - e\gamma_{\mu}A_{\mu} \tag{4.135}$$

which can be written as

$$\not{\partial} \to \not{\partial} - ie\not{A} \tag{4.136}$$

and the Dirac equation with the electromagnetic interaction included becomes

$$(\not{\partial} - ie\not{A} + m)\psi = 0 \tag{4.137}$$

We shall see that this elegant and seemingly simple equation contains an incredible amount of physical information. It is seen to be covariant.

To find the solution of Equation (4.137) we follow the same path as we followed in the beginning of Section 4.3. We look for plane wave solutions of the form Equation (4.78) and we find instead of Equation (4.85)

$$\begin{pmatrix} m - (E - e\phi) & \boldsymbol{\sigma}\cdot(\mathbf{p} - e\mathbf{A}) \\ -\boldsymbol{\sigma}\cdot(\mathbf{p} - e\mathbf{A}) & m + (E - e\phi) \end{pmatrix}\begin{pmatrix} u_A \\ u_B \end{pmatrix} = 0 \tag{4.138}$$

As always, $\mathbf{A}$ and ϕ are the vector and scalar potentials of the electromagnetic field from which the electric and magnetic fields can be calculated. Equation (4.138) could have been obtained directly from Equation (4.85) when the substitution in Equation (4.133) is applied directly to Equation (4.85), but we wanted to display the Dirac equation with the electromagnetic interaction included, see Equation (4.137). To obtain non-trivial solutions of the set of linear and homogeneous equations such as Equation (4.138), we calculated the determinant and required it to be zero. Here care must be taken when doing that because the operator $\mathbf{p}$ will act on $\mathbf{A}$ and ϕ if $\mathbf{p}$ is to their left, so their order matters. Instead of calculating the determinant, we solve Equation (4.138) by the method of substitution. We solve for u_B in terms of u_A from the second equation in Equation (4.138) and substitute the resulting u_B in the first equation of Equation (4.138). Refer to the discussion below Equation (4.98) for an explanation of this choice. We obtain u_B from the second equation in Equation (4.138) by multiplying that equation from the left by the inverse of $m + (E - e\phi)$

$$u_B = \frac{1}{m + (E - e\phi)} \boldsymbol{\sigma} \cdot (\mathbf{p} - e\mathbf{A})\, u_A \tag{4.139}$$

Note the order of the two factors in Equation (4.139); $\mathbf{p}$ differentiates ϕ so do not exchange these factors. Substitution of u_B of Equation (4.139) in the first equation of Equation (4.138) gives

$$[m - (E - e\phi)]u_A + \boldsymbol{\sigma} \cdot (\mathbf{p} - e\mathbf{A}) \frac{1}{m + (E - e\phi)} \boldsymbol{\sigma} \cdot (\mathbf{p} - e\mathbf{A}) u_A = 0 \tag{4.140}$$

Anticipating taking the non-relativistic limit we set

$$E = m + E' \quad \text{with} \quad E' \ll m \tag{4.141}$$

We have in Equation (4.140) the factors

$$m - (E - e\phi) = -E' + e\phi \tag{4.142}$$

$$m + (E - e\phi) = 2m + E' - e\phi = 2m\left(1 + \frac{E' - e\phi}{2m}\right) \tag{4.143}$$

so

$$\frac{1}{m + (E - e\phi)} = \frac{1}{2m(1 + \frac{E' - e\phi}{2m})} = \frac{1}{2m}\left(1 - \frac{E' - e\phi}{2m} + \ldots\right) \tag{4.144}$$

where we have used that

$$\frac{1}{1 + x} = 1 - x + \cdots \quad \text{for} \quad |x| \ll 1 \tag{4.145}$$

so we required in Equation (4.144) that

$$\frac{E' - e\phi}{2m} \ll 1 \tag{4.146}$$

We will discuss this condition below. Introducing Equation (4.142) and Equation (4.144) in Equation (4.140) gives

$$(-E' + e\phi) + \boldsymbol{\sigma} \cdot (\mathbf{p} - e\mathbf{A}) \frac{1}{2m} \left(1 - \frac{E' - e\phi}{2m} + \ldots \right) \boldsymbol{\sigma} \cdot (\mathbf{p} - e\mathbf{A}) = 0 \tag{4.147}$$

We want to solve for E', but E' appears in two places in Equation (4.147). The first term is of $O(v^2)$ while the second term contains a term of $O(v^2)$ and one of $O(v^4)$. This is so because without the $(E' - e\phi)/(2m)$ term in the second term, Equation (4.147) corresponds to the Schrödinger equation for which all terms are of $O(v^2)$ (the potential energy is of the same order as the kinetic energy). Thus the $(E' - e\phi)/(2m)$ term in the second term makes that part of the second term $O(v^4)$. The expectation value of the velocity of the electron in a hydrogen atom (in units of the velocity of light) is approximately $\alpha \approx 1/137$ so the neglected term is of $O(\alpha^2/4) \approx 10^{-5}$. This approximation is a safe one. When we neglect $(E' - e\phi)/(2m)$ in the second term in Equation (4.147) we obtain

$$(-E' + e\phi) + [\boldsymbol{\sigma} \cdot (\mathbf{p} - e\mathbf{A})]^2 \frac{1}{2m} = 0 \tag{4.148}$$

Because $(\boldsymbol{\sigma} \cdot \mathbf{a})^2 = \mathbf{a}^2$, see Equation (4.88), one might evaluate $[\boldsymbol{\sigma} \cdot (\mathbf{p} - e\mathbf{A})]^2$ to be $(\mathbf{p} - e\mathbf{A})^2$. Because $\mathbf{p}$ and $\mathbf{A}$ do not commute, we have to be more careful here. To see this, check the derivation of the relation $(\boldsymbol{\sigma} \cdot \mathbf{a})^2 = \mathbf{a}^2$. We use instead the relation in Equation (4.87) to obtain the general relation

$$(\boldsymbol{\sigma} \cdot \mathbf{a})(\boldsymbol{\sigma} \cdot \mathbf{b}) = \mathbf{a} \cdot \mathbf{b} + i\boldsymbol{\sigma} \cdot (\mathbf{a} \times \mathbf{b}) \tag{4.149}$$

Thus we get

$$\begin{aligned} [\boldsymbol{\sigma} \cdot (\mathbf{p} - e\mathbf{A})]^2 &= (\mathbf{p} - e\mathbf{A})^2 + i\boldsymbol{\sigma} \cdot [\mathbf{p} \times \mathbf{p} - e\,\mathbf{p} \times \mathbf{A} - e\,\mathbf{A} \times \mathbf{p} + e^2\mathbf{A} \times \mathbf{A}] \\ &= (\mathbf{p} - e\mathbf{A})^2 - ie\,\boldsymbol{\sigma} \cdot (\mathbf{p} \times \mathbf{A} + \mathbf{A} \times \mathbf{p}) \end{aligned} \tag{4.150}$$

But $\mathbf{p} \times \mathbf{A}\,\psi = \mathbf{p}_A \times \mathbf{A}\,\psi + \mathbf{p}_\psi \times \mathbf{A}\,\psi = (1/i)(\nabla \times \mathbf{A})\,\psi - \mathbf{A} \times \mathbf{p}\,\psi$ where we used the product rule of differentiation and ψ is a test function that depends upon position. We have exchanged the $\mathbf{p}_\psi$ and $\mathbf{A}$ factors to put $\mathbf{p}_\psi$ last so that its subscript ψ can be dropped. Therefore $\mathbf{p} \times \mathbf{A} = (1/i)\nabla \times \mathbf{A} - \mathbf{A} \times \mathbf{p}$. The last term in parenthesis in Equation (4.150) becomes $\mathbf{p} \times \mathbf{A} + \mathbf{A} \times \mathbf{p} = (1/i)\nabla \times \mathbf{A} = (1/i)\mathbf{B}$ where the

$-\mathbf{A} \times \mathbf{p}$ and the $\mathbf{A} \times \mathbf{p}$ terms have canceled and $\mathbf{B} = \nabla \times \mathbf{A}$ is the magnetic field. So Equation (4.150) becomes

$$[\sigma \cdot (\mathbf{p} - e\mathbf{A})]^2 = (\mathbf{p} - e\mathbf{A})^2 - ie\,\sigma \cdot \left(\frac{1}{i}\mathbf{B}\right) = (\mathbf{p} - e\mathbf{A})^2 - e\,\sigma \cdot \mathbf{B} \tag{4.151}$$

We return to Equation (4.148), substitute Equation (4.151) in it, and solve for E' to get

$$E' = \frac{1}{2m}(\mathbf{p} - e\mathbf{A})^2 + e\phi - \frac{e}{2m}\sigma \cdot \mathbf{B} \tag{4.152}$$

With the identification of E' with the Hamiltonian H we recover the Schrödinger equation with an extra term

$$H' = -\frac{e}{2m}\sigma \cdot \mathbf{B} \tag{4.153}$$

This extra term corresponds to the potential energy $-\boldsymbol{\mu} \cdot \mathbf{B}$ of a magnetic moment $\boldsymbol{\mu}$ in a magnetic field $\mathbf{B}$ that is constant over the volume of themagnetic moment with $\boldsymbol{\mu}$ given by

$$\boldsymbol{\mu} = \frac{e}{2m}\sigma = \frac{e}{2m}2\mathbf{S} = g\frac{e}{2m}\mathbf{S} \tag{4.154}$$

The minus sign is part of the classical expression of the potential energy of a magnetic moment in a magnetic field that is constant over the volume of the magnetic moment. We see from Equation (4.154) that the g-factor of a Dirac particle is 2, independent of its mass. This is a great success for Dirac's theory. This solves the puzzle encountered earlier in the development of Quantum Physics, namely that the magnetic moment of an electron is two Bohr magnetons $e/(2m)$ instead of one as might be naïvely expected.

This is not the whole story. When second quantization principles are applied to the Dirac theory, using the second quantized electromagnetic field we arrive at Quantum Electrodynamics (QED). QED improved upon the calculation of the magnetic moment, leading to a more precise expression for the g-factor. We find for example for the electron or positron

$$g_e = 2\left(1 + \frac{\alpha}{2\pi} - 0.328\left(\frac{\alpha}{2\pi}\right)^2 + \ldots\right) = 2\,(1 + \kappa_e) \tag{4.155}$$

where α is the finestructure constant and κ_e is called the anomalous part of the g-factor leading to the anomalous part of the magnetic moment of an electron or positron. The first two terms in the first parentheses in Equation (4.155) are obviously independent of the mass of the Dirac particle, but the next terms depend upon its mass, hence the subscript e on g and κ. Experiment gives 0.327 ± 0.005 for the coefficient in the third term in Equation (4.155) in excellent agreement with theory. The g-factors of the electron and positron have been measured to be equal to better than 4 parts in 10^{12}. The Dirac theory predicts them to be exactly equal if the mass of the electron and the positron are equal. There exists in nature a 'heavy electron' called the muon. There are positive and negative muons. Its mass is 105.7 MeV compared with the electron mass of 0.511 MeV. Because the mass of the electron and muon are different, their g-factors are predicted by QED to be different as well, due to the third and following terms in the first parentheses in Equation (4.155). The magnetic moment of the muon has also been measured and was found to agree with the QED prediction. The difference between the g-factors of the positive and negative muons is measured to be smaller than 5 parts in 10^8.

It is interesting to note that the proton and the neutron have very large anomalous magnetic moments. The proton has a measured anomalous part of the g-factor equal to $\kappa_p = 1.79$ when the magnetic moment is expressed in nuclear Bohr magnetons $(e/(2m_N))$ with m_N the nucleon mass. The neutron's magnetic moment should be zero in the Dirac theory but it is measured to be equal to $-1.91(e/(2m_N)$, entirely anomalous. The proton and neutron are spin $\frac{1}{2}$ particles so they should satisfy the Dirac equation. They obviously do not, so we may conclude that they are not elementary particles. Indeed, it was discovered in the 1970s that they have quarks and gluons as constituents. There are indications that the quarks themselves satisfy the Dirac equation, that is, they appear to have the correct magnetic moment. This topic is outside the scope of the current discussion.

We know that a particle with orbital angular momentum $\mathbf{L}$ has a magnetic moment $\mu = -e/(2m)\mathbf{L}$. Thus the complete term in the Hamiltonian from the magnetic moments due to the orbital and intrinsic angular momenta is

$$H' = -\frac{e}{2m}(\mathbf{L} + g\mathbf{S}) \cdot \mathbf{B} \tag{4.156}$$

4.5 THE HYDROGEN ATOM RE-REVISITED

To study the hydrogen atom we take the non-relativistic limit of the Dirac equation as in the previous section, and set $\mathbf{A} = 0$. We obtain, using Equation (4.147)

$$(-E' + U) + \boldsymbol{\sigma} \cdot \mathbf{p}\frac{1}{2m}\left(1 - \frac{E' - U}{2m} + \ldots\right)\boldsymbol{\sigma} \cdot \mathbf{p} = 0 \tag{4.157}$$

where we have set $e\phi = U$, the potential energy. This time we keep the $(E' - e\phi)/(2m)$ term but neglect the higher order terms beyond it. We solve for E' and get

$$E' = U + \frac{\mathbf{p}^2}{2m} - \boldsymbol{\sigma} \cdot \mathbf{p}\frac{E' - U}{4m^2}\boldsymbol{\sigma} \cdot \mathbf{p} \tag{4.158}$$

The term on the left and the first two terms on the right are of $O(v^2)$ while the last term is of $O(v^4)$. This is so because without the last term Equation (4.158) corresponds to the Schrödinger equation for which the terms are of $O(v^2)$ (the potential energy U is of the same order as the kinetic energy). The last term has two powers of $\mathbf{p}$ and a factor of the same order as the other terms in Equation (4.158), making it $O(v^4)$. But v is of order α, see the discussion following Equation (4.147), so the last term in Equation (4.158) is very small relative to the other terms. Unfortunately E' appears on both sides in Equation (4.158), but because the last term is so small we can get an approximate solution for E' from Equation (4.158) by neglecting the last term and substituting that estimate back into the last term of Equation (4.158). We find

$$E' - U \approx \frac{\mathbf{p}^2}{2m} \tag{4.159}$$

and E' will be gone from the right-hand side of Equation (4.158). We would like to commute the last two of the three factors in the last term of Equation (4.158) so that we have $(\boldsymbol{\sigma} \cdot \mathbf{p})^2(E' - U)$ which evaluates to $\mathbf{p}^2(E' - U)$. Because $E' - U$ does not commute with $\boldsymbol{\sigma} \cdot \mathbf{p}$, we introduce their commutator and write

$$(E' - U)\boldsymbol{\sigma} \cdot \mathbf{p} = \boldsymbol{\sigma} \cdot \mathbf{p}\,(E' - U) + [E' - U, \boldsymbol{\sigma} \cdot \mathbf{p}] \tag{4.160}$$

and substitute the estimate Equation (4.159) for $E' - U$ on the right-hand side of Equation (4.160). This introduces an error of $O(v^5) = O(\alpha^5)$ in a term that is of $O(v^3) = O(\alpha^3)$, see the discussion following Equation (4.158). In the commutator $[E' - U, \boldsymbol{\sigma} \cdot \mathbf{p}]$, E' commutes with $\boldsymbol{\sigma} \cdot \mathbf{p}$ so the commutator becomes $[-U, \boldsymbol{\sigma} \cdot \mathbf{p}] = \boldsymbol{\sigma} \cdot [\mathbf{p}, U]$ and Equation (4.160) becomes

$$(E' - U)\boldsymbol{\sigma} \cdot \mathbf{p} \approx \boldsymbol{\sigma} \cdot \mathbf{p}\frac{\mathbf{p}^2}{2m} + \boldsymbol{\sigma} \cdot [\mathbf{p}, U] \tag{4.161}$$

We now multiply the left and right side of Equation (4.161) by $\boldsymbol{\sigma} \cdot \mathbf{p}$ and divide by $4m^2$ to obtain an expression for the last term in Equation (4.158)

$$\begin{aligned} \boldsymbol{\sigma} \cdot \mathbf{p}\frac{E' - U)}{4m^2}\boldsymbol{\sigma} \cdot \mathbf{p} &= \frac{(\boldsymbol{\sigma} \cdot \mathbf{p})^2}{4m^2}\frac{\mathbf{p}^2}{2m} + \frac{\boldsymbol{\sigma} \cdot \mathbf{p}}{4m^2}\boldsymbol{\sigma} \cdot [\mathbf{p}, U] \\ &= \frac{\mathbf{p}^4}{8m^3} + \frac{1}{4m^2}\mathbf{p} \cdot [\mathbf{p}, U] + \frac{i}{4m^2}\boldsymbol{\sigma} \cdot \mathbf{p} \times [\mathbf{p}, U] \end{aligned} \tag{4.162}$$

For the last term we used Equation (4.149) with $\mathbf{a} = \mathbf{p}$ and $\mathbf{b} = [\mathbf{p}, U]$. Substitution of Equation (4.162) into Equation (4.158) and solving for E' gives

$$E' = U + \frac{\mathbf{p}^2}{2m} - \frac{\mathbf{p}^4}{8m^3} - \frac{i\boldsymbol{\sigma} \cdot \mathbf{p} \times [\mathbf{p}, U]}{4m^2} - \frac{\mathbf{p} \cdot [\mathbf{p}, U]}{4m^2} \tag{4.163}$$

The $\mathbf{p}^4$ term is a relativistic correction and can be derived also by using the Einstein relation as follows

$$E' = E - m = \sqrt{\mathbf{p}^2 + m^2} - m = m\sqrt{1 + \frac{\mathbf{p}^2}{m^2}} - m \tag{4.164}$$

Using the expansion $\sqrt{1+\varepsilon} = 1 + \frac{1}{2}\varepsilon - \frac{1}{8}\varepsilon^2 + \ldots$ in Equation (4.164) we obtain

$$E' = m\left(1 + \frac{\mathbf{p}^2}{2m^2} - \frac{\mathbf{p}^4}{8m^4} + \ldots\right) - m = \frac{\mathbf{p}^2}{2m} - \frac{\mathbf{p}^4}{8m^3} + \ldots \tag{4.165}$$

compare with Equation (4.163). The fourth term on the right-hand side of Equation (4.163) is called the Thomas term and can be evaluated as

$$-\frac{i\boldsymbol{\sigma}\cdot\mathbf{p}\times[\mathbf{p}, U]}{4m^2} = -\frac{\boldsymbol{\sigma}\cdot\mathbf{p}\times\nabla U}{4m^2} \tag{4.166}$$

$$= -\frac{e^2}{4m^2r^3}\boldsymbol{\sigma}\cdot(\mathbf{p}\times\mathbf{r}) \tag{4.167}$$

$$= \frac{e^2}{4m^2r^3}\boldsymbol{\sigma}\cdot(\mathbf{r}\times\mathbf{p}) \tag{4.168}$$

$$= \frac{e^2}{2m^2r^3}\mathbf{S}\cdot\mathbf{L} \tag{4.169}$$

For the first equality in Equation (4.166) we used that $[\mathbf{p}, U]\psi = (1/i)[\nabla, U]\psi = (1/i)(\nabla U - U\nabla)\psi = (1/i)(\nabla_U U + \nabla_\psi U - U\nabla)\psi = (1/i)(\nabla U)\psi$ where ψ is a test function and we used the product rule for the derivative of a product. The symbol ∇_F means that only the function F will be differentiated. For the second equality Equation (4.167) we set $U = -e^2/r$, the Coulomb potential for the hydrogen atom, and use that $\nabla U = e^2\,\mathbf{r}/r^3$. To obtain the third equality Equation (4.168) we used that $\mathbf{p}\times\mathbf{r} = -\mathbf{r}\times\mathbf{p}$ even though $\mathbf{p}$ and $\mathbf{r}$ do not commute, because we have, using again a test function ψ, that $(\mathbf{p}\times\mathbf{r})\,\psi = (1/i)(\nabla\times\mathbf{r})\,\psi = (1/i)(\nabla_\mathbf{r}\times\mathbf{r} + \nabla_\psi\times\mathbf{r})\,\psi = -\mathbf{r}\times\nabla_\psi\psi$ because $\nabla\times\mathbf{r} = 0$. In the last equality we used that $\mathbf{S} = \boldsymbol{\sigma}/2$. The result in Equation (4.169) is called the spin-orbit coupling because it involves the dot product of the spin and the orbital angular momentum. The factor 2 in the prefactor in Equation (4.169) is called the Thomas factor, after Thomas who derived it using special relativistic considerations. Here the factor 2 appears automatically. The nature of the spin-orbit coupling becomes clear if we consider that the electric field $\mathbf{E}$ of the hydrogen nucleus at the position of the electron is given by $\mathbf{E} = -\nabla U$, so the right-hand side of Equation (4.166) is proportional to $\boldsymbol{\sigma}\cdot(\mathbf{v}\times\mathbf{E}) = \boldsymbol{\sigma}\cdot\mathbf{B}$. So the term corresponds to the potential energy of the magnetic moment of the electron in the presence of the magnetic field generated by the motion of the electron through the Coulomb field of the hydrogen nucleus. One may do this calculation by naïvely considering a coordinate system attached to the electron and observing the hydrogen nucleus circling the electron. The moving hydrogen nucleus corresponds to a circular electric current which sets up a magnetic field by Ampere's law. The magnetic moment thus acquires a potential energy. Because the coordinate system so defined is not an inertial coordinate system, one cannot expect to get the correct answer. Thomas did the calculation correctly and found that the naïve calculation is off by a factor of 2.

The last term in Equation (4.163) can be evaluated using $\mathbf{p} \cdot [\mathbf{p}, U] \propto \boldsymbol{\nabla} \cdot (\boldsymbol{\nabla} U) \propto \boldsymbol{\nabla} \cdot \mathbf{E} = -4\pi\, |e|\, \delta^3(\mathbf{r})$ to get

$$\frac{\mathbf{p} \cdot [\mathbf{p}, U]}{4m^2} = \frac{\pi\, e^2}{2m^2}\, \delta^3(\mathbf{r}) \tag{4.170}$$

Because of the $\delta^3(\mathbf{r})$ dependence, the result in Equation (4.170) is called the contact term because it contributes only if the electron and hydrogen nucleus are at the same spatial point. It is also called the Darwin term. Therefore it affects only *s*-states because only these have wavefunctions that are non-zero at $\mathbf{r} = 0$. For example

$$\langle n, 0, 0 | H_D | n, 0, 0 \rangle = \frac{m\alpha^4}{2n^3} \tag{4.171}$$

Note the dependence on α^4 as expected from the considerations following Equation (4.158).

Our calculation of E' is based upon an expansion in powers of v^2. An exact result can be obtained

$$E(n, j) = m\left[1 + \left(\frac{\alpha}{n - (j + \frac{1}{2}) + \sqrt{(j + \frac{1}{2})^2 - \alpha^2}}\right)^2\right]^{-\frac{1}{2}} \tag{4.172}$$

Here n is the principal quantum number and j the eigenvalue of the total angular momentum operator $\mathbf{J} = \mathbf{L} + \mathbf{S}$. All states with given values of n and j are degenerate to all orders of α. One can expand the result in Equation (4.172) in powers of α as

$$E(n, j) = m\left[1 - \frac{\alpha^2}{2n^2} - \frac{\alpha^4}{2n^3}\left(\frac{1}{j + \frac{1}{2}} - \frac{3}{4n}\right) + O(\alpha^6)\right] \tag{4.173}$$

Even though the eigenvalues of $\mathbf{L}$ are not good quantum numbers, they are still used in the spectroscopic notation for energy levels. For example $1s_{1/2}$ means $n = 1$, $l = 0$, and $j = \frac{1}{2}$. It is seen from Equation (4.172) that the states $2s_{1/2}$ and $2p_{1/2}$ are still degenerate as they were in the non-relativistic treatment of the hydrogen atom.

In the late 1940s Lamb and Retherford observed experimentally that that degeneracy is ever so slightly lifted by about 1060 MHz with the 2*s* higher energy level. This is in contradiction with the exact calculation using the Dirac equation. The solution of this contradiction is resolved in a QED based calculation. It gives 1057.7 ± 0.2 MHz while the experimental value is 1057.8 ± 0.1 MHz, in excellent agreement with QED.

PROBLEMS

(1) *Dirac Algebra.*
  (a) Prove that $[\gamma_i, \gamma_j] = 2i\varepsilon_{ijk}\Sigma_k$ and $\gamma_i\gamma_j = \delta_{ij} + i\varepsilon_{ijk}\Sigma_k$.
  Define the chirality operator $\gamma_5 = \gamma_1\gamma_2\gamma_3\gamma_4$.
  (b) Prove that γ_5 is Hermitian and that $\{\gamma_\mu, \gamma_5\} = 2\delta_{\mu\nu}, \mu, \nu = 1, 5$. Prove this without using an explicit representation for γ_5 but by using the properties of the γ_μ.
  (c) Calculate γ_5 and verify the statements in (b).
  (d) Prove that $\Sigma_k = -(i/2)[\gamma_i, \gamma_j]$ $(i, j, k = \text{cyclic})$ where Σ_k represents the spin.
  (e) Prove that $\Sigma_k = i\gamma_4\gamma_5\gamma_k$.
  (f) Calculate the expectation value of the velocity dx_k/dt using the Dirac Hamiltonian, first without and then with the electromagnetic interaction included.
  (g) Using the result from (e), calculate the square of the expectation value of the velocity. Comment on the result.

(2) *Dirac Spinors.*
  The Dirac equation is linear and homogeneous. This means that its solutions are determined up to a multiplicative factor N.
  (a) Prove this statement.
  (b) Find N such that $\bar{u}_r(\mathbf{p})u_r(\mathbf{p}) = -\bar{v}_s(\mathbf{p})v_s(\mathbf{p}) = 1$, $r = 1, 2, 3, 4$ and $s = 1, 2$. No summation over repeated indices is implied.
  (c) Show that $\bar{u}_r(\mathbf{p})u_s(\mathbf{p}) = \delta_{rs}$ for $r, s = 1, 2$.
  (d) Show that $\bar{v}_r(\mathbf{p})v_s(\mathbf{p}) = -\delta_{rs}$ for $r, s = 1, 2$.
  (e) Show that $\bar{u}_r(\mathbf{p})v_s(\mathbf{p}) = \delta_{rs}$ and $\bar{v}_r(\mathbf{p})u_s(\mathbf{p}) = \delta_{rs}$ for $r, s = 1, 2$.

(3) *Probability Current.*
  The probability current was shown to be $j_\mu = i\bar{\psi}\gamma_\mu\psi$.
  (a) Prove that $\mathbf{j} = \mathbf{v}$ with $\mathbf{v}$ the particle's velocity, and prove that $j_4 = iE/m$. Use the normalization factor N found in Problem 2(b).
  (b) Comment on your answer in (a) and show in particular that the E/m factor in j_4 makes $\int \psi^\dagger \psi d^3\mathbf{x}$ Lorenz invariant.
  (c) Show that the probability current is conserved even in the presence of an electromagnetic field.

(4) *Massless Dirac Particles.*

(a) Write down the Dirac equation for a massless particle and redo the derivation of the four u-spinors and the two v-spinors.

(b) Prove that $hu_i = -\gamma_5 u_i$ for $i = 1,2$ and $hu_i = +\gamma_5 u_i$ for $i = 3,4$ where the u_i are spinors for massless Dirac particles and h is the helicity operator $h = \mathbf{\Sigma} \cdot \mathbf{p} / \mid \mathbf{p} \mid$.

(c) Prove that $hv_i = -\gamma_5 v_i$ for $i = 1,2$ where the v_i are spinors for massless Dirac particles.

(d) Using the results from (b) and (c), cast the equations for the four u-spinors and the two v-spinors in the form $(1 \pm \gamma_5)$spinor $= 0$.

(e) The results from (b) and (c) show that for massless Dirac particles the helicity operator and the chirality operator are equal to each other up to a sign. Is this also true for massive Dirac particles?

(5) Show that

$$\overline{u}\gamma_\mu u = \frac{p_\mu}{im}\overline{u}u \qquad \text{and} \qquad \overline{v}\gamma_\mu v = -\frac{p_\mu}{im}\overline{v}v$$

5

Special Topics

5.1 INTRODUCTION

Quantum Physics involves concepts that are thought by some to be contrary to intuition. In particular the notion of interfering amplitudes and probability associated with physical phenomena, and the act of a measurement leading to the collapse of the wave function, have led some to question the validity of Quantum Physics. There is a large body of literature where these matters are discussed in detail. To illustrate these concerns we will discuss some examples in the sections that follow.

Human intuition is shaped by observing macroscopic objects whose scales are matched to the abilities of the sensory organs of humans. Quantum Physics, on the other hand, deals with microscopic phenomena that require the use of macroscopic instruments for their observation. Therefore these microscopic phenomena contribute to our intuition to a much lesser extent. Rather than relying upon intuition, a large majority of physical scientists hold the view that, as long as Quantum Physics is in agreement with experimental observations, it stands.

5.2 MEASUREMENTS IN QUANTUM PHYSICS

We first review the concept of measurements in Quantum Physics. It is assumed that a discussion of this topic has already taken place in an earlier course in Quantum Physics. We therefore briefly summarize the concepts and refer to an introductory text for a more detailed discussion.

A measurement of a quantity represented by the operator Q results in one of the eigenvalues of the operator Q. We note those eigenvalues by $\langle Q \rangle_i$. Reproducibility of measurements requires that if we repeat the

An Introduction to Advanced Quantum Physics Hans P. Paar

same measurement on the same system we get the same eigenvalue. This forces the notion of the 'collapse of the wave function'.

Suppose that the state vector of a system is $|\psi\rangle$. This state vector can be expanded in a series of eigenkets $|i\rangle$ of an operator Q as

$$|\psi\rangle = \sum_i a_i|i\rangle \tag{5.1}$$

For $|\psi\rangle$ to be normalized, we require that

$$\sum_i |a_i|^2 = 1 \tag{5.2}$$

Quantum Physics states that a measurement of a property of the system represented by the operator Q results in an eigenvalue of Q, say $\langle Q\rangle_i$ with probability $\mathcal{P}_i = |a_i|^2$. Thus a first measurement gives

$$\langle Q\rangle_i = \langle i|Q|i\rangle \tag{5.3}$$

For the measurement to be reproducible, a subsequent measurement must give the same result with probability $\mathcal{P}_i = 1$. Thus $a_i = 1$ which, because of Equation (5.2), implies that all other a_j $(j \neq i)$ are zero. This is the 'collapse of the wave function', a result of the first measurement.

We discuss next an example that involves a spin $\frac{1}{2}$ particle. The reader is assumed to be thoroughly familiar with the spin gymnastics that are needed to understand this system. Suppose we have a system consisting of a single spin $\frac{1}{2}$ particle whose spin is aligned with the positive z-axis. This can be expressed as

$$S_z|+z\rangle = \tfrac{1}{2}|+z\rangle \tag{5.4}$$

where we set $|\frac{1}{2},\frac{1}{2}\rangle = |+z\rangle$ and $\hbar = 1$. The kets $|+z\rangle$ and $|-z\rangle$ form a complete set and we used one of these to define the state of the system. The kets $|+x\rangle$ and $|-x\rangle$ also form a complete set and we might have used these instead to define the system. Of course the eigenkets relative to the z-axis can be expressed in terms of the eigenkets relative to the x-axis. We know from our studies of spin $\frac{1}{2}$ systems that

$$|+z\rangle = \frac{|+x\rangle + |-x\rangle}{\sqrt{2}} \tag{5.5}$$

To see this, use that

$$|+z\rangle = \cos\tfrac{1}{2}\theta\,|+\rangle + e^{i\phi}\sin\tfrac{1}{2}\theta\,|-\rangle \tag{5.6}$$

where θ specifies the direction of a new axis of quantization and is measured relative to the z-axis, and the kets $|+\rangle$ and $|-\rangle$ are relative to that new axis.

We now make a measurement of the spin direction along the x-axis, that is, we evaluate $S_x|+z\rangle$ with $|+z\rangle$ given by Equation (5.5). The outcome is either $+\frac{1}{2}$ or $-\frac{1}{2}$ with equal probability as the coefficients of $|+x\rangle$ and $|-x\rangle$ are equal. The outcome cannot be predicted. We know that this is a consequence of the fact that the commutator $[S_x, S_z] \neq 0$.

Suppose that the measurement represented by S_x yielded $-\frac{1}{2}$. The state vector $|+z\rangle$ has collapsed (is reduced) to one term as a result of the measurement, namely the second term in Equation (5.5). When one repeats the measurement, one obtains the same $-\frac{1}{2}$ answer as before, thus guaranteeing reproducibility. If one now measures the direction of the spin along the z-axis, that is, we evaluate $S_z|-x\rangle$, it is found that the outcome is either $+\frac{1}{2}$ or $-\frac{1}{2}$ with equal probability because

$$|-x\rangle = \frac{|+z\rangle - |-z\rangle}{\sqrt{2}} \tag{5.7}$$

and the outcome cannot be predicted. Compare this outcome with the original situation specified by Equation (5.4).

We now apply this to the Einstein-Podolsky-Rosen paradox.

5.3 EINSTEIN-PODOLSKY-ROSEN PARADOX

Einstein, Podolsky and Rosen formulated what is called the EPR Paradox. Perhaps calling it the EPR experiment is a more suitable name; we shall simply call it 'the EPR'. The proposal is an ingenious one as one might expect when one of the founders of Quantum Physics is involved. The proposal can be formulated in different ways, but all involve the collapse of the wave function by making two observations in locations far from each other and within a time interval so short that there is no causal connection between them. EPR formulated their experiment in terms of non-commuting variables (momentum and position). It was later formulated in terms of spin.

Our formulation of the EPR involves the decay of a spin 0 parent particle into two photons. Another example would be the decay of a spin 0 parent particle into two spin $\frac{1}{2}$ particles which we consider in Section 5.5. The outcome of either illustrates the EPR. Examples of the former can be found in Particle Physics where, for example, the parent particle is a π^0, a particle with spin 0 and mass of approximately 135.0 MeV/c^2. Its lifetime of approximately 8.4×10^{-17} s is in agreement with the expectation of an electromagnetic decay $\pi^0 \rightarrow \gamma\gamma$, consistent with the presence of the two photons in the final state. As discussed in Section 1.6, the electromagnetic interaction conserves parity. Indeed the intrinsic parity of the π^0 is well defined and has been measured to be negative, and we expect the parity of the two photon state to be negative as well. The particular eigenvalue of the parity operator does not affect the essence of the EPR discussion.

We will analyze the $\pi^0 \to \gamma\gamma$ in the π^0 center-of-mass. To conserve momentum, the photons must move in opposite directions with momenta **k** and −**k** respectively. We can analyze the spin of the photons by labeling their kets either by their circular polarization ($|R\rangle$ and $|L\rangle$) or by their polarization ($|x\rangle$ and $|y\rangle$). Here x and y are two mutually perpendicular directions that are both perpendicular to the momenta of the photons which are taken to be parallel to the z-axis. When the photons have left the region of their production, they do not interact and their state can be represented by the product of their respective kets. Thus the state with two photons can be written for example as $|+\vec{k}, R\rangle|-\vec{k}, L\rangle$ or $|+\vec{k}, x\rangle|-\vec{k}, y\rangle$ where the first ket applies to photon 1 and the second to photon 2. From here on we drop the momentum from the kets and write the above example product kets as $|R\rangle|L\rangle = |R_1, L_2\rangle$ and $|x\rangle|y\rangle = |x_1, y_2\rangle$. The kets $|R\rangle$ and $|L\rangle$ of a single photon form a complete orthonormal set, as do the kets $|x\rangle$ and $|y\rangle$

$$\langle R|L\rangle = \langle L|R\rangle = 0 \tag{5.8}$$

$$\langle R|R\rangle = \langle L|L\rangle = 1 \tag{5.9}$$

$$\langle x|y\rangle = \langle y|x\rangle = 0 \tag{5.10}$$

$$\langle x|x\rangle = \langle y|y\rangle = 1 \tag{5.11}$$

Each pair can be expressed in terms of the other pair as

$$|R\rangle = \frac{|x\rangle + i|y\rangle}{\sqrt{2}} \tag{5.12}$$

$$|L\rangle = \frac{|x\rangle - i|y\rangle}{\sqrt{2}} \tag{5.13}$$

and their inverse relations

$$|x\rangle = \frac{|R\rangle + i|L\rangle}{\sqrt{2}} \tag{5.14}$$

$$|y\rangle = \frac{|R\rangle - i|L\rangle}{i\sqrt{2}} \tag{5.15}$$

To conserve angular momentum in the decay $\pi^0 \to \gamma\gamma$, the two photons must either be both right-handed or both left-handed, that is, they can be either in the state $|R_1, R_2\rangle$ or $|L_1, L_2\rangle$. Orbital angular momentum does not contribute along the z-axis (classically, it is perpendicular to the z-axis). States such as $|R_1, L_2\rangle$ are forbidden by angular momentum conservation. The most general state vector for the two photons can be written as a linear superposition of $|R_1, R_2\rangle$ and $|L_1, L_2\rangle$

$$|\psi\rangle = p|R_1, R_2\rangle + q|L_1, L_2\rangle \tag{5.16}$$

To normalize $|\psi\rangle$ we require that

$$|p|^2 + |q|^2 = 1 \tag{5.17}$$

Applying the parity operator to for example $|R\rangle$ we get

$$P|R\rangle = a|L\rangle \qquad\qquad P|L\rangle = b|R\rangle \tag{5.18}$$

because under the parity operator the momentum of a photon is reversed but its spin is not (see the discussion of parity in Section 1.6). The prefactors a and b in these two equations must satisfy $a^2 = b^2 = 1$, as can be seen by applying the parity operator twice to either $|R\rangle$ or $|L\rangle$ and using $P^2 = 1$ and Equation (5.18). Thus we take $a = b = 1$ in Equation (5.18), as a phase will not be observable.

Applying the parity operator to $|\psi\rangle$ we get

$$P|\psi\rangle = P[p\,|R_1, R_2\rangle + q\,|L_1, L_2\rangle] = p\,|L_1, L_2\rangle + q\,|R_1, R_2\rangle \tag{5.19}$$

For this to equal $-|\psi\rangle = -p|R_1, R_2\rangle - q|L_1, L_2\rangle$ we must require that $p = -q$ because the kets $|R_1, R_2\rangle$ and $|L_1, L_2\rangle$ are orthonormal. Because the overall sign of ψ has no physical meaning and taking into account the normalization condition Equation (5.17), we set $p = -q = 1/\sqrt{2}$. Thus Equation (5.16) becomes

$$|\psi\rangle = \frac{|R_1, R_2\rangle - |L_1, L_2\rangle}{\sqrt{2}} \tag{5.20}$$

The two photons are detected by two observers A and B who have traveled to their collinear and remote positions while taking care to keep their clocks synchronized (they are physicists who took a course in Special Relativity) and to keep the x and y axis of their respective coordinate systems aligned. They are at nearly equal distances from the location of the π^0. We assume that observer A detects photon 1 and observer B photon 2. The observers carry instruments that allow them to measure the linear polarization perpendicular to the momentum of the photon they detect. The observers measure the polarization of photons they detect, the time of detection and note these. They observe large numbers of π^0 decays. After some time they meet and compare notes. The amplitude for observer A to find his photon polarized along the x-axis and observer B to find his photon to be polarized along the x-axis is given by

$$\langle x_1, x_2|\psi\rangle = \frac{\langle x_1, x_2|R_1, R_2\rangle - \langle x_1, x_2|L_1, L_2\rangle}{\sqrt{2}} = 0 \tag{5.21}$$

where we used that

$$\langle x|R\rangle = \frac{1}{\sqrt{2}} \qquad \langle y|R\rangle = \frac{i}{\sqrt{2}} \tag{5.22}$$

$$\langle x|L\rangle = \frac{1}{\sqrt{2}} \qquad \langle y|L\rangle = -\frac{i}{\sqrt{2}} \tag{5.23}$$

as can be obtained from Equation (5.12) and Equation (5.13). One can also show that

$$\langle x_1, y_2|\,\psi\rangle = \frac{i}{\sqrt{2}} \tag{5.24}$$

$$\langle y_1, x_2|\,\psi\rangle = \frac{i}{\sqrt{2}} \tag{5.25}$$

$$\langle y_1, y_2|\,\psi\rangle = 0 \tag{5.26}$$

We note at this point that the minus sign in Equation (5.20) is of no importance, it merely switches the values of the amplitudes in Equation (5.21) and Equation (5.24) through Equation (5.26). The interpretation of these results is as follows (we assume that observer A makes his measurement slightly before observer B makes his):

(1) Observer A and observer B each detect left-handed and right-handed photons. They observe these two polarizations in equal numbers. This is due to the fact that the two terms in Equation (5.20) have the same magnitude.
(2) If observer A detects a right-handed photon, observer B also detects a right-handed photon. If observer A detects a left-handed photon, observer B also detects a left-handed photon. In Quantum Physics this is due to the collapse of the wave function in Equation (5.20) due to the measurement of the photon's polarization by observer A. More about this in the next section.
(3) The EPR motivated critic of Quantum Physics has no problem with these facts because (1) is a consequence of the fact that the polarization of the emitted photons is random and (2) is a consequence of the conservation of angular momentum. The critic assumes that the two photons are emitted in either the state $|R_1, R_2\rangle$ or $|L_1, L_2\rangle$ with equal probability, and the observers merely determine which is the case in each π^0 decay.
(4) If observer A chooses to measure his photon's polarization along the x-direction then observer B has a finite probability to find that the polarization of his photon is along the y-direction. This follows from Equation (5.24) which is based upon Equation (5.20). Observer A can change the orientation of his polarimeter after the π^0 decay has taken place, guaranteeing that the outcome of observer A's measurement

is randomly distributed between $|x\rangle$ and $|y\rangle$. Yet the measurement of B depends upon the outcome of the measurement of A, even though when A does not make his measurement, the outcome of B's measurement is equally probable $|x\rangle$ or $|y\rangle$.

(5) If observer A chooses to measure his photon's polarization along the x-direction then observer B has zero probability to find that the polarization of his photon is also along the x-direction. This follows from Equation (5.21) which is based upon Equation (5.20). The same remarks can be made as in the previous case.

(6) The critic cannot explain items (4) and (5) using angular momentum conservation.

(7) The EPR motivated critic rejects the correlations mentioned as 'instantaneous action at a distance' because the outcome of the type of measurement chosen by observer A appears to affect the outcome of the measurement by observer B, even though the photons are far apart when they are detected.

The items (4) and (5) above are indeed remarkable and contrary to most people's intuition, unless that intuition is shaped by Quantum Physics. Yet experiments show that the correlations embodied in Equation (5.21) and Equation (5.24) through Equation (5.26) are correct. We say that the two photons are entangled, that is, a measurement of a property of one affects the outcome of the measurement of a property of the other. Entanglement is not limited to photons, other systems such as two spin $\frac{1}{2}$ particles that result from the decay of a spin 0 parent are also entangled, and the above discussion applies to that system as well. Entanglement provides a method of transferring messages between two parties whose content cannot be obtained by a third party without destroying the message's content.

5.4 SCHRÖDINGER'S CAT

The notion of the collapse of a state vector as the result of a measurement has been questioned repeatedly. One particular way of doing that is called Schrödinger's Cat. It involves a thought experiment in which a cat is locked in a box that contains a window that is normally blocked. In the box is also a radioactive source. When the source emits a particle at a random time a poison will be released that kills the cat. An observer outside the box can open the window and look inside the box. The cat is considered to be either alive or dead so the ket describing the state of the cat is written as

$$|\text{cat}\rangle = a|\text{alive}\rangle + b|\text{dead}\rangle \tag{5.27}$$

where the coefficients a and b are time dependent and satisfy $|a|^2 + |b|^2 = 1$. At $t = 0$ we have that $a = 1$ and $b = 0$. After some time, $a < 1$ while $b > 0$.

The observer outside the box will take the state vector of the cat to still be given by Equation (5.27) but he would like to know whether the cat is dead or alive. To answer this question he decides to open the window and have a look. According to Quantum Physics this amounts to a measurement and the state vector Equation (5.27) must collapse according to what we discussed in Section 5.2. If enough time has passed it is likely that the observer will observe the cat to be dead, while before the measurement the cat had a finite amplitude to be alive. The act of looking in the box will have some likelihood to kill the cat.

This is obviously an incorrect application of Quantum Physics, but why is it incorrect? The answer should be supplied by Quantum Physics itself. Unfortunately this is not the case but instead we reason as follows. The origin of the problem must be found in the fact that the cat is a macroscopic object that contains many elementary systems (molecules). Those elementary systems can each be described by state vectors and the cat as a whole must be described by the product of such state vectors. Because Avogadros's number is very very large, these product vectors contain a huge number of factors and there are a huge number of different product vectors. The cat as a whole cannot be described by a state vector that is a superposition of just two terms.

This 'explanation' however begs the question as to how many elementary systems an object may have before the application of Quantum Physics becomes incorrect. Because the answer to that question is not known, we draw an analogy with the situation in Quantum Electrodynamics. There we encountered a number operator (whose eigenvalue specifies the number of photons) and electric and magnetic field operators, see the discussion in Section 1.2 near the end. It was shown that the number operator's eigenvalue on the one hand, and the electric and magnetic fields on the other, are not simultaneously observable with arbitrary precision because the number operator does not commute with the electric and magnetic field operators. We do not let that stand in the way of discussing systems with a small number of photons by methods based upon Quantum Electrodynamics (ignoring the fact that electric and magnetic fields are not well defined) or a discussion of the signal from a radio station by methods based upon classical electromagnetism involving electric and magnetic fields (ignoring the fact that the number operator's eigenvalues are not well defined). We did not specify precisely how the applicability of quantum versus classical methods depends upon the number of photons in the system. In the case of a macroscopic object like the cat we call upon the judgement of the scientist to pick the correct method for analyzing the system without giving precise guidance as to how many elementary systems are allowed in the makeup of the system.

This is not a very healthy situation. We have applied Quantum Physics to a proton and yet the proton contains a large number of constituents:

various species ('flavors') of quarks and anti-quarks as well as gluons. This does not invalidate the applicability of Quantum Physics to a proton.

A proper scientific theory should be able to specify its range of applicability. Perhaps in the future such a specification will be found. For the moment it appears that one must rely upon one's good judgement.

5.5 THE WATCHED POT

Another attempt to attack the notion of the collapse of the state vector by a measurement is provided by 'The Watched Pot' or 'A watched pot will not boil over'. This is also called the Zeno Effect after the fourth century Greek philosopher who was well known for inventing paradoxes. Misra and Sudarshan published on this effect in 1977, but earlier papers already allude to this effect. As a pot is heated from a low temperature it is initially not boiling over. As time progresses the state vector of the pot is a linear superposition of not boiling over and boiling over. When an observer looks at the pot, its state vector collapses to one of the two terms in the linear superposition. Because the pot initially is not boiling over, if the observer looks often enough at the pot during the heating process, the state vector collapses each time he looks at the not boiling over state.

Let us analyze the situation within the formalism of time-dependent perturbation theory. A system that has a probability $P(0)$ to be in the state $|i\rangle$ at $t = 0$ will after some time t be found in that same state with probability

$$P(t) = P(0)e^{-t/\tau} \tag{5.28}$$

where t is the time of a first measurement and τ is the inverse of the transition probability away from the state $|i\rangle$. In what follows we will let $t \to 0$ so $t \ll \tau$ and neglect terms of $\mathcal{O}(t/\tau)^2$. Thus we may approximate Equation (5.28) as

$$P(t) = P(0)\left(1 - \frac{t}{\tau}\right) \tag{5.29}$$

We note that in some derivations Equation (5.28) is derived from Equation (5.29).

If instead we make a first measurement at $2t$ instead of t we find that the probability for the system to be in the state $|i\rangle$ is

$$P(t) = P(0)\left(1 - \frac{2t}{\tau}\right) \tag{5.30}$$

where we used Equation (5.29). This outcome is consistent with the outcome of making two measurements, one at t and one at $2t$, because the probability to find the system in the state $|i\rangle$ after the first measurement is

$P(t) = P(0)(1 - t/\tau)$ and thus to find it in the state $|i\rangle$ after the second measurement is $P(t)(1 - t/\tau)$ where we used Equation (5.29) twice, the second time replacing $P(0)$ by $P(t)$. Thus we find

$$P(2t) = (1 - t/\tau) \cdot (1 - t/\tau) = (1 - t/\tau)^2 = \left(1 - \frac{2t}{\tau}\right) \tag{5.31}$$

In general, making n measurements with $n \to \infty$ in a finite time T, the interval between successive measurements is $t = T/n$ and we find for the probability to find the system in the state $|i\rangle$ at time T

$$P(T) = P(0)(1 - t/\tau)^n = 1 - T/\tau \tag{5.32}$$

Here we used the binomial expansion

$$(1 + a)^n = 1 + na \quad \text{for } a \to 0 \tag{5.33}$$

So far so good, as far as Quantum Physics is concerned, but let's look in detail at the condition $t \to 0$. In the derivation of Equation (5.28) and Equation (5.29) for the transition from an initial state $|i\rangle$ to a final state $|f\rangle$ under the influence of a harmonic perturbation, we encountered the result in Equation (1.95)

$$\dot{c}_f = \kappa\, \mathrm{e}^{i(E_B - E_A + \omega)t} \tag{5.34}$$

see Section 1.3. Here c_f is the amplitude of the final state $|f\rangle$ and κ is a constant whose form does not matter for the present discussion. E_A and E_B are the initial and final state energies of the system and ω is the energy of an emitted photon. The expression in Equation (5.34) was integrated over time from $-\infty$ to $+\infty$ to get an expression proportional to $\delta(E_B - E_A + \omega)$. The δ-function expresses energy conservation. In the present situation we integrate over time from 0 (when the perturbation is turned on) to t to obtain

$$c_f(t) = \kappa \int_0^t \mathrm{e}^{i(E_B - E_A + \omega)t'}\, dt' = \kappa\, \frac{\mathrm{e}^{i(E_B - E_A + \omega)t} - 1}{E_B - E_A + \omega} \tag{5.35}$$

Squaring this result to obtain the probability to find the system in the state $|f\rangle$ we find

$$|c_f(t)|^2 = \kappa^2\, \frac{4 \sin^2\left[\frac{1}{2}(E_B - E_A + \omega)t\right]}{(E_B - E_A + \omega)^2} = \kappa^2 t^2 \quad \text{for } t \to 0 \tag{5.36}$$

Because we consider a system with only two possible states $|i\rangle$ and $|f\rangle$ we have that $|c_i(t)|^2 = 1 - |c_f(t)|^2$ and thus

$$P(t) = |c_i(t)|^2 = 1 - |c_f(t)|^2 = 1 - \kappa^2 t^2 \tag{5.37}$$

Compare with Equation (5.29), $P(0) = 1$ in the present case. The t^2 dependence causes trouble as can be seen as follows. Following the reasoning below Equation (5.29), making a first measurement at $2t$ instead of t, we find using Equation (5.37)

$$P(2t) = 1 - \kappa^2(2t)^2 = 1 - 4\kappa^2 t^2 \tag{5.38}$$

If we make two measurements, one at t and one at $2t$ we would find, using the same reasoning as before

$$P(2t) = (1 - \kappa^2 t^2)^2 = 1 - 2\kappa^2 t^2 \tag{5.39}$$

again using the binomial expansion in Equation (5.33). Making n measurements with $n \to \infty$ in a finite time T we find

$$P(T) = \left[1 - \kappa^2\left(\frac{T}{n}\right)^2\right]^n = 1 - \frac{\kappa^2 T^2}{n} \to 1 \quad \text{for } n \to \infty \tag{5.40}$$

This result appears to confirm the 'Watched Pot Does Not Boil Over' hypothesis. The results Equation (5.38) and Equation (5.39) are inconsistent and the critic would say that this is due to the first observation's collapse of the state vector. The inconsistency indicates that Quantum Physics predicts that the rate of decay of the system depends upon the frequency of observations when the observations are spaced by infinitely short time intervals. Does this indicate that Quantum Physics has a problem? It is thought not because it is in principle impossible to observe a system at infinitely short time intervals and draw sensible conclusions. Heisenberg's Uncertainty relation indicates that the system's energy becomes uncertain as we shorten the time interval between observations. Defining such time intervals requires that the boundaries of each time interval must be defined to a precision at least as good as the length of the time interval, that is, with near zero uncertainty. Thus the arguments that lead to Equation (5.40) are untenable and the result in Equation (5.40) is incorrect.

5.6 HIDDEN VARIABLES AND BELL'S THEOREM

Some critics of Quantum Physics have addressed the questions raised by the EPR by postulating that Quantum Physics is correct as far as it goes, but that it is incomplete and that this is the reason for the outcomes of measurements to be probabilistic rather than deterministic. It is postulated that 'hidden variables' exist that precisely determine the outcome of measurements. Because we do not know about those variables, the outcome of measurements appear to have a probabilistic character.

Bell in 1964 published a deep analysis of the EPR and the possible role of hidden variables. We consider again a π^0 decay but this time not into two photons but into a back-to-back electron and positron: $\pi^0 \rightarrow e^+ e^-$. The two observers A and B of Section 5.3 are equipped to measure the direction of the spins of the electron and positron along the direction of the unit vectors $\hat{\mathbf{a}}$ and $\hat{\mathbf{b}}$ respectively. These unit vectors are both perpendicular to the momenta of the electron and positron respectively. Bell introduced the correlation between the measurements of observers A and B as follows. $A(\hat{\mathbf{a}})$ is the spin measured by observer A along the direction $\hat{\mathbf{a}}$. The outcome of A's measurement is either $A(\hat{\mathbf{a}}) = +\hbar/2$ or $A(\hat{\mathbf{a}}) = -\hbar/2$. A similar situation holds for observer B. The correlation coefficient C is defined as

$$C(\hat{\mathbf{a}}, \hat{\mathbf{b}}) = \frac{4}{\hbar^2} < A(\hat{\mathbf{a}})B(\hat{\mathbf{b}}) > \tag{5.41}$$

The correlation coefficient can take on values $-1 \leq C \leq 1$ owing to the factor $4/\hbar^2$ in Equation (5.41). $C = 0$ if the outcomes of the measurements of A and B are uncorrelated because in that case each spin is measured to be $\pm\hbar/2$ with random $\pm$ signs independent of each other. If $\hat{\mathbf{a}} = \hat{\mathbf{b}}$, that is the two spins are measured along the same direction, the spins must point in opposite directions because the total angular momentum must be zero (the π^0 is a spin 0 particle and angular momentum is conserved in its decay) so

$$A(\hat{\mathbf{a}}) = -B(\hat{\mathbf{a}}) \tag{5.42}$$

and thus

$$C(\hat{\mathbf{a}}, \hat{\mathbf{a}}) = -1 \tag{5.43}$$

Likewise $C(\hat{\mathbf{a}}, -\hat{\mathbf{a}}) = +1$. Quantum Physics predicts that

$$C = -\hat{\mathbf{a}} \cdot \hat{\mathbf{b}} = -\cos\theta \tag{5.44}$$

where θ is the angle between $\hat{a}$ and $\hat{b}$, see Problem 5(2).

Bell's seminal observation was that the result in Equation (5.44) is impossible in a Quantum Theory with hidden variables. He reasoned as follows. Assume that the outcome of the measurement of the electron's spin is independent of the measurement of the positron's spin. This is reasonable because each observer can change the direction of his polarimeter just prior to the arrival of the particle of which he is measuring the spin. Let the hidden variable be λ, that is, λ determines whether $A(\hat{\mathbf{a}}, \lambda)$ equals $+\hbar/2$ or $-\hbar/2$ in a manner that we do not have to specify. Similarly $B(\hat{\mathbf{b}}, \lambda) = \pm\hbar/2$ with the sign determined by λ. Similar to Equation (5.42) we have

$$A(\hat{\mathbf{a}}, \lambda) = -B(\hat{\mathbf{a}}, \lambda) \tag{5.45}$$

for all λ. We introduce a probability density $\rho(\lambda)$ of which λ is of course normalized

$$\int \rho(\lambda)d\lambda = 1 \tag{5.46}$$

with $\rho(\lambda)d\lambda$ is the probability to find a value of λ in the interval $[\lambda, \lambda + d\lambda]$. We use it to define the correlation coefficient as

$$C(\hat{\mathbf{a}}, \hat{\mathbf{b}}) = \frac{4}{\hbar^2}\int \rho(\lambda)A(\hat{\mathbf{a}}, \lambda)B(\hat{\mathbf{b}}, \lambda)d\lambda \tag{5.47}$$

similar to Equation (5.41). We now use Equation (5.45) by replacing $\hat{\mathbf{a}}$ by $\hat{\mathbf{b}}$ in it, to eliminate $B(\hat{\mathbf{b}}, \lambda)$ in Equation (5.47) to get

$$C(\hat{\mathbf{a}}, \hat{\mathbf{b}}) = -\frac{4}{\hbar^2}\int \rho(\lambda)A(\hat{\mathbf{a}}, \lambda)A(\hat{\mathbf{b}}, \lambda)d\lambda \tag{5.48}$$

If $\hat{\mathbf{c}}$ is yet another unit vector pointing in its own direction we can use Equation (5.48) to get

$$C(\hat{\mathbf{a}}, \hat{\mathbf{c}}) = -\frac{4}{\hbar^2}\int \rho(\lambda)A(\hat{\mathbf{a}}, \lambda)A(\hat{\mathbf{c}}, \lambda)d\lambda \tag{5.49}$$

We now take the difference $C(\hat{\mathbf{a}}, \hat{\mathbf{b}}) - C(\hat{\mathbf{a}}, \hat{\mathbf{c}})$ using Equation (5.48) and Equation (5.49)

$$C(\hat{\mathbf{a}}, \hat{\mathbf{b}}) - C(\hat{\mathbf{a}}, \hat{\mathbf{c}}) = -\frac{4}{\hbar^2}\int \rho(\lambda)[A(\hat{\mathbf{a}}, \lambda)A(\hat{\mathbf{b}}, \lambda) - A(\hat{\mathbf{a}}, \lambda)A(\hat{\mathbf{c}}, \lambda)]d\lambda \tag{5.50}$$

$$= -\int \rho(\lambda)\left[1 - \frac{4}{\hbar^2}A(\hat{\mathbf{b}}, \lambda)A(\hat{\mathbf{c}}, \lambda)\right]\left[\frac{4}{\hbar^2}A(\hat{\mathbf{a}}, \lambda)A(\hat{\mathbf{b}}, \lambda)\right]d\lambda \tag{5.51}$$

To derive Equation (5.51) from Equation (5.50) we took $A(\hat{\mathbf{a}}, \lambda)A(\hat{\mathbf{b}}, \lambda)$ outside parentheses and used that $A(\hat{\mathbf{b}}, \lambda)A(\hat{\mathbf{b}}, \lambda) = \hbar^2/4$. Because $-1 \leq (4/\hbar^2)A(\hat{\mathbf{b}}, \lambda)A(\hat{\mathbf{c}}, \lambda) \leq +1$ the first bracketed expression in Equation (5.51) is positive or zero, and because the probability density $\rho(\lambda)$ is positive or zero, $\rho(\lambda)[1 - (4/\hbar^2)A(\hat{\mathbf{b}}, \lambda)A(\hat{\mathbf{c}}, \lambda)]$ is positive or zero. The second bracketed expression in Equation (5.51) likewise satisfies $-1 \leq (4/\hbar^2)A(\hat{\mathbf{a}}, \lambda)A(\hat{\mathbf{b}}, \lambda) \leq +1$, so if we leave it out of the integrand that makes the integrand larger or leaves it the same, so we conclude that

$$|C(\hat{\mathbf{a}}, \hat{\mathbf{b}}) - C(\hat{\mathbf{a}}, \hat{\mathbf{c}})| \leq \int \rho(\lambda)\left[1 - \frac{4}{\hbar^2}A(\hat{\mathbf{b}}, \lambda)A(\hat{\mathbf{c}}, \lambda)\right] \tag{5.52}$$

$$\leq 1 + C(\hat{\mathbf{b}}, \hat{\mathbf{c}}) \tag{5.53}$$

This is Bell's inequality. Note that we did not make any assumptions about $\rho(\lambda)$ except that it must satisfy the conditions for a probability density.

One can show that Bell's inequality is inconsistent with Equation (5.44), which we assume to hold in a Quantum Theory with hidden variables because we assumed that the standard Quantum Theory is incomplete, not that it is wrong. Assume that $\hat{\mathbf{a}}$ and $\hat{\mathbf{b}}$ are perpendicular to each other and that $\hat{\mathbf{c}}$ is their bisectrice. According to Equation (5.44) we have $C(\hat{\mathbf{a}}, \hat{\mathbf{b}}) = 0$, $C(\hat{\mathbf{a}}, \hat{\mathbf{c}}) = -1/\sqrt{2}$, and $C(\hat{\mathbf{b}}, \hat{\mathbf{c}}) = -1/\sqrt{2}$. Substitution in Equation (5.53) gives

$$\left|0 - \left(-\frac{1}{\sqrt{2}}\right)\right| \leq 1 + \left(-\frac{1}{\sqrt{2}}\right) \tag{5.54}$$

and this is an incorrect inequality!

So Quantum Theory, supplemented with hidden variables introduced to address the EPR, is internally inconsistent. Either one or the other of the following is true: (i) Quantum Physics without hidden variables is correct and the EPR is not a paradox, or (ii) Quantum Physics without hidden variables is wrong (not incomplete but wrong) and there is no EPR paradox! Rather than let our 'intuition' decide between these two possibilities, we should ask what nature (experiment) has to say about this. Numerous increasingly sophisticated experiments overwhelmingly select possibility (i). In one experiment the experimenters went as far as changing randomly the orientation of the polarimeters after the emission of the two particles (photons in their case), and they used the fiber optic network of the telephone company in Geneva, Switzerland, to propagate the photons over many kilometers.

We note that causality is not violated because the collapse of the wave function cannot be used to transmit information instantaneously between the two observers. We know of another case where it appears that velocities greater than the speed of light are possible: group velocity in propagation of waves. Also there we do not have a violation of causality.

To conclude we quote Feynman's statement about the various questions that have been raised about the validity of Quantum Physics:

> 'We have always had a great deal of difficulty in understanding the world-view that quantum mechanics represents. At least, I do.... It has not yet become obvious to me that there is no real problem. I cannot define the real problem, therefore I suspect there's no real problem, but I'm not sure there's no real problem.'

PROBLEMS

(1) *The EPR.*

(a) Show that $|R_1R_2\rangle - |L_1L_2\rangle = i[|x_1y_2\rangle + |y_1x_2\rangle]$.

(b) Connect this result with the discussion in Section 5.3.

(2) *Derivation of* $C(\hat{a}, \hat{b}) = -\cos\theta$ *of Equation (5.44).*

(a) Rephrase Equation (5.44) in terms of Pauli matrices.

(b) Find the relation between $\boldsymbol{\sigma}_1$ and $\boldsymbol{\sigma}_2$ for the singlet state of two spin $\frac{1}{2}$ particles.

(c) Evaluate $(\sigma_1 \cdot \hat{\mathbf{a}})(\sigma_2 \cdot \hat{\mathbf{b}})$, and use (b) to eliminate σ_2.

(d) Calculate the correlation coefficient $C(\hat{\mathbf{a}}, \hat{\mathbf{b}})$.

Part II

Introduction to Quantum Field Theory

6

Second Quantization of Spin 1/2 and Spin 1 Fields

6.1 SECOND QUANTIZATION OF SPIN $\frac{1}{2}$ FIELDS

6.1.1 Plane Wave Solutions

In Chapter 1 we applied second quantization to the electromagnetic field. It resulted in the description of processes in which photons were created or destroyed. The results were in excellent agreement with theory. When the creation of electron-positron pairs was observed with the discovery of the positron in 1932 by Anderson, it became clear that second quantization of the Dirac theory of electrons and their anti-particles, the positrons, was called for. The formalism is based upon work by Jordan and Wigner on anti-commuting creation and annihilation operators in 1928.

In analogy with the procedure followed for the photon, we consider a free electron described by the Dirac equation. Because the four spinors $u_r(\mathbf{p})$ $(r = 1, 4)$ form a complete set and the Dirac equation is linear and homogeneous, general solutions of the Dirac equation for a free particle can be written as a linear superposition of four plane waves as

$$\psi(x) = \frac{1}{\sqrt{V}} \sum_{\mathbf{p},r} a_r(\mathbf{p}) u_r(\mathbf{p})\, e^{ipx} \tag{6.1}$$

where V is the volume of the space in which we quantize the physical system; m, $E_p = \pm\sqrt{\mathbf{p}^2 + m^2}$, and $\mathbf{p}$ are the mass, energy and momentum, and $a_r(\mathbf{p})$ $(r = 1, 4)$ are the amplitudes belonging to each spinor $u_r(\mathbf{p})$. The first two terms in the sum over r correspond to a particle with positive energy $E_p > 0$ while the last two terms correspond to a particle with negative energy $E_p < 0$. If we are going to interpret $a_r(\mathbf{p})$ as an annihilation operator that annihilates a spin $\frac{1}{2}$ particle, then the last two terms correspond to the annihilation of a particle with negative energy. This may be interpreted as the creation of an anti-particle with positive energy: compare with the discussion in Section 4.3. Thus we should replace the annihilation operators $a_3(\mathbf{p})$ and $a_4(\mathbf{p})$ and spinors $u_3(\mathbf{p})$ and $u_4(\mathbf{p})$ by creation operators and spinors for anti-particles $b_2^\dagger(\mathbf{p})$ and $b_1^\dagger(\mathbf{p})$ and $v_2(\mathbf{p})$ and $v_1(\mathbf{p})$ respectively. The Hermitian conjugate $\dagger$ notation instead of the complex conjugate $*$ notation anticipates the notation of second quantization of the Dirac equation. We must also replace $\mathbf{p} \to -\mathbf{p}$ and $E_p \to -E_p$. Refer to Equation (4.105) and Equation (4.106) for the choice of the indices 1 and 2 on v. Thus Equation (6.1) becomes

$$\psi(x) = \frac{1}{\sqrt{V}} \sum_{\mathbf{p},r} \sqrt{\frac{m}{E_p}} \left[a_r(\mathbf{p}) u_r(\mathbf{p}) e^{ipx} + b_r^\dagger(\mathbf{p}) v_r(\mathbf{p}) e^{-ipx} \right] \tag{6.2}$$

where now $r = 1, 2$ and $E_p = \sqrt{\mathbf{p}^2 + m^2} \geq 0$; compare with Section 1.1 for the photon. The prefactor $1/\sqrt{V}$ and the factor $\sqrt{m/E_p}$ merely redefine the operators $a_r(\mathbf{p})$ and $b_r^\dagger(\mathbf{p})$. It is clear that $\psi(x)$ is not Hermitian.

Taking the Hermitian conjugate and multiplying from the right by γ_4 we obtain

$$\overline{\psi}(x) = \frac{1}{\sqrt{V}} \sum_{\mathbf{p},r} \sqrt{\frac{m}{E_p}} \left[a_r^\dagger(\mathbf{p}) \overline{u}_r(\mathbf{p}) e^{-ipx} + b_r(\mathbf{p}) \overline{v}_r(\mathbf{p}) e^{ipx} \right] \tag{6.3}$$

where

$$\overline{u} = u^\dagger \gamma_4 \qquad \overline{v} = v^\dagger \gamma_4 \tag{6.4}$$

in analogy with the definition $\overline{\psi} = \psi^\dagger \gamma_4$, see Equation (4.67). We anticipate that if a and $b^\dagger$ annihilate particles and create anti-particles of energy E_p and momentum $\mathbf{p}$, that then $a^\dagger$ and b create particles and annihilate anti-particles of energy E_p and momentum $\mathbf{p}$. Here E_p is positive in all four cases because of the introduction of v_1 and v_2 instead of u_4 and u_3. The factor $\sqrt{m/E_p}$ in Equation (6.2) and Equation (6.3) is chosen such that the energy of the second quantized field is sensible, as will be shown below. The factor depends upon the normalization of the spinors u and v. Their normalization has been the subject of Problem 2 of Chapter 4.

6.1.2 Normalization of Spinors

Using the expressions Equation (4.101) and Equation (4.107) for the spinors u and v, one can show that $(r, s = 1, 2)$

$$u_r^\dagger(\mathbf{p}) u_s(\mathbf{p}) = \delta_{rs} N^2 \frac{2E_p}{E_p + m} \tag{6.5}$$

$$v_r^\dagger(\mathbf{p})v_s(\mathbf{p}) = \delta_{rs}N^2\frac{2E_p}{E_p+m} \tag{6.6}$$

$$\bar{u}_r(\mathbf{p})u_s(\mathbf{p}) = \delta_{rs}N^2\frac{2m}{E_p+m} \tag{6.7}$$

$$\bar{v}_r(\mathbf{p})v_s(\mathbf{p}) = -\delta_{rs}N^2\frac{2m}{E_p+m} \tag{6.8}$$

Note the minus sign in Equation (6.8).

The normalization constant N that appears in Equation (4.101) and Equation (4.107) and therefore in Equation (6.5) through Equation (6.8) is usually chosen such that the probability $\psi^\dagger\psi\, d^3\mathbf{x}$ to find a particle in a certain three-dimensional volume is Lorenz invariant. The three-dimensional volume $d^3\mathbf{x}$ in any inertial frame is related to the three-dimensional volume $d^3\mathbf{x}_0$ in the rest system of the particle by $d^3\mathbf{x} = d^3\mathbf{x}_0/\gamma = d^3\mathbf{x}_0\, m/E$. Therefore we compensate for the contraction of the volume by normalizing $\psi^\dagger\psi = u^\dagger u = v^\dagger v$ to $\gamma = E/m$. In the rest system of the particle this factor equals 1 as it should. As can be seen from Equation (6.5) through Equation (6.8) this normalization is accomplished by choosing $N^2 = (E_p+m)/(2m)$. With this value of N, relations Equation (6.5) through Equation (6.8) become $(r,s = 1,2)$

$$u_r^\dagger(\mathbf{p})u_s(\mathbf{p}) = \delta_{rs}\frac{E_p}{m} \tag{6.9}$$

$$v_r^\dagger(\mathbf{p})v_s(\mathbf{p}) = \delta_{rs}\frac{E_p}{m} \tag{6.10}$$

$$\bar{u}_r(\mathbf{p})u_s(\mathbf{p}) = \delta_{rs} \tag{6.11}$$

$$\bar{v}_r(\mathbf{p})v_s(\mathbf{p}) = -\delta_{rs} \tag{6.12}$$

Note the minus sign in Equation (6.12).

Using the expressions Equation (4.101) and Equation (4.107) for the spinors u and v and the normalization constant N discussed above, one can also show that $(r,s = 1,2)$

$$u_r^\dagger(\mathbf{p})v_s(\mathbf{q}) = \varepsilon_{rs}\left[\frac{\boldsymbol{\sigma}\cdot\mathbf{p}}{2m} + \frac{\boldsymbol{\sigma}\cdot\mathbf{q}}{2m}\right] \tag{6.13}$$

$$v_r^\dagger(\mathbf{p})u_s(\mathbf{q}) = \varepsilon_{rs}\left[\frac{\boldsymbol{\sigma}\cdot\mathbf{p}}{2m} + \frac{\boldsymbol{\sigma}\cdot\mathbf{q}}{2m}\right] \tag{6.14}$$

$$\bar{u}_r(\mathbf{p})v_s(\mathbf{q}) = -\varepsilon_{rs}\left[\frac{\boldsymbol{\sigma}\cdot\mathbf{p}}{2m} - \frac{\boldsymbol{\sigma}\cdot\mathbf{q}}{2m}\right] \tag{6.15}$$

$$\bar{v}_r(\mathbf{p})u_s(\mathbf{q}) = -\varepsilon_{rs}\left[\frac{\boldsymbol{\sigma}\cdot\mathbf{p}}{2m} - \frac{\boldsymbol{\sigma}\cdot\mathbf{q}}{2m}\right] \tag{6.16}$$

where ε_{rs} is antisymmetric in r and s with $\varepsilon_{12} = 1$. Equations (6.14) and (6.16) are the Hermitian conjugates of Equations (6.13) and (6.15) respectively.

For future reference we consider the following special cases. We set $\mathbf{q} = -\mathbf{p}$ in Equation (6.13) and Equation (6.14) and obtain $(r, s = 1, 2)$

$$u_r^\dagger(\mathbf{p}) v_s(-\mathbf{p}) = 0 \tag{6.17}$$

$$v_r^\dagger(\mathbf{p}) u_s(-\mathbf{p}) = 0 \tag{6.18}$$

We can also set $\mathbf{q} = \mathbf{p}$ in Equation (6.15) and Equation (6.16) and obtain $(r, s = 1, 2)$

$$\overline{u}_r(\mathbf{p}) v_s(\mathbf{p}) = 0 \tag{6.19}$$

$$\overline{v}_r(\mathbf{p}) u_s(\mathbf{p}) = 0 \tag{6.20}$$

6.1.3 Energy

We now calculate the energy of a free Dirac particle described by the plane wave solutions Equation (6.2) and Equation (6.3). The energy E is given by the expectation value of the Hamiltonian $H = \boldsymbol{\alpha} \cdot \mathbf{p} + \beta m$, see Equation (4.32), with $\alpha_i = i\gamma_4\gamma_i$ and $\beta = \gamma_4$, see Equation (4.33).

$$H = \int d^3\mathbf{x}\, \psi^\dagger H \psi \tag{6.21}$$

$$= \int d^3\mathbf{x}\, \overline{\psi}\gamma_4 H \psi \tag{6.22}$$

$$= \int d^3\mathbf{x}\, \overline{\psi}\gamma_4 (i\gamma_4\gamma_i p_i + \gamma_4 m)\psi \tag{6.23}$$

$$= \int d^3\mathbf{x}\, \overline{\psi}(i\gamma_i p_i + m)\psi \tag{6.24}$$

$$= \int d^3\mathbf{x}\, \overline{\psi}(\gamma_i \partial_i + m)\psi \tag{6.25}$$

Using the Dirac Equation (4.27), the expression $(\gamma_i\partial_i + m)\psi$ can be written as $-\gamma_4\partial_4\psi$ and Equation (6.26) becomes

$$E = -\int d^3\mathbf{x}\, \overline{\psi}\gamma_4\partial_4\psi \tag{6.26}$$

It is noteworthy that this is the expectation value of $-(\hbar/i)\partial/\partial t$, the operator that corresponds to the energy, see Equation (4.2). Substitution of Equation (6.2) and Equation (6.3) gives

$$E = -\int \frac{d^3\mathbf{x}}{V} \sum_{\mathbf{p},\mathbf{p}',r,r'} \frac{m}{\sqrt{E_p E_p'}} \left[a_r^\dagger(\mathbf{p})\overline{u}_r(\mathbf{p})e^{-ipx} + b_r(\mathbf{p})\overline{v}e^{ipx}\right]\gamma_4\, ip_4' \\ \left[a_{r'}(\mathbf{p}')u_{r'}(\mathbf{p}')e^{ip'x} - b_{r'}^\dagger(\mathbf{p}')v_{r'}(\mathbf{p}')e^{-ip'x}\right] \qquad (6.27)$$

Because $p_4' = iE_p'$ we have that $ip_4' = -E_p'$ and this minus sign cancels the minus sign in front of the integral. The factor γ_4 between the two factors in square brackets will cancel the γ_4 in $\overline{u} = u^\dagger\gamma_4$ and $\overline{v} = v^\dagger\gamma_4$ leaving $u^\dagger$ and $v^\dagger$. The sum is evaluated in much the same manner as the sum in the expression for $|\mathbf{E}|^2$ in Equation (1.46). After multiplying out the two factors with square brackets, we obtain four terms with exponentials of the form $\exp[\pm i(p - p')x]$ and $\exp[\pm i(p + p')x]$. We exchange the order of the integration and the summation and do the integral first using Equation (1.48). Thus the double sum over $\mathbf{p}, \mathbf{p}'$ reduces to a single sum over $\mathbf{p}$ and we have two types of terms: those with $\mathbf{p}' = \mathbf{p}$ and those with $\mathbf{p}' = -\mathbf{p}$. In the former the time dependences in the exponentials cancel, while in the latter we get factors with exponentials of the form $\exp(-2iEt)$ and $\exp(+2iEt)$. The latter two terms are proportional to $u^\dagger(\mathbf{p})v(-\mathbf{p})$ and $v^\dagger(\mathbf{p})u(-\mathbf{p})$ respectively, so they are zero by virtue of the relations Equation (6.17) and Equation (6.18). The former two terms are proportional to $u_r^\dagger(\mathbf{p})u_{r'}(\mathbf{p})$ and $v_r^\dagger(\mathbf{p})v_{r'}(\mathbf{p})$ and these can be evaluated using Equation (6.9) and Equation (6.10) and are seen to equal $(E_p/m)\delta_{rr'}$. Therefore the double sum over r and r' reduces to a single sum over r and the expression for E becomes

$$E = \sum_{\mathbf{p},r} \frac{m}{E_p}\left[a_r^\dagger(\mathbf{p})\, a_r(\mathbf{p}) - b_r(\mathbf{p})\, b_r^\dagger(\mathbf{p})\right]\frac{E_p}{m} E_p \qquad (6.28)$$

$$= \sum_{\mathbf{p},r}\left[a_r^\dagger(\mathbf{p})\, a_r(\mathbf{p}) - b_r(\mathbf{p})\, b_r^\dagger(\mathbf{p})\right]E_p \qquad (6.29)$$

This expression is independent of time as it should be for a free particle. Note the order of the product of $a^\dagger, a$ on the one hand and $b, b^\dagger$ on the other as well as the minus sign. In Equation (6.28) the factor m/E_p comes from the $\sqrt{(m/E_p)}$ in Equations (6.2) and (6.3), the factor E_p/m comes from the normalization of the spinors u and v, and the factor E_p comes from the factor p_4' between the two factors in square brackets in Equation (6.27).

We require the energy to be positive definite because we have replaced the negative energy solutions for particles by positive energy solutions for anti-particles. Yet the expression in Equation (6.29) is not positive definite because of the minus sign. At this point in the discussion of the energy of the electromagnetic field, we introduced commutation relations between a and $a^\dagger$ leading to the annihilation and creation operators for photons and the occupation number operator $N = a^\dagger a$. In the present situation we need to introduce two occupation number operators, one for particles ($N^{(1)}$) and one for anti-particles ($N^{(2)}$)

$$N_r^{(1)}(\mathbf{p}) = a_r^\dagger(\mathbf{p})\, a_r(\mathbf{p}) \qquad (r = 1, 2) \tag{6.30}$$

$$N_r^{(2)}(\mathbf{p}) = b_r^\dagger(\mathbf{p})\, b_r(\mathbf{p}) \qquad (r = 1, 2) \tag{6.31}$$

To eliminate the minus sign in Equation (6.29) we postulate anticommutation relations instead of commutation relations ($r, r' = 1, 2$)

$$\{a_r(\mathbf{p}), a_{r'}^\dagger(\mathbf{p}')\} = \delta_{\mathbf{pp}'}\, \delta_{rr'} \tag{6.32}$$

$$\{b_r(\mathbf{p}), b_{r'}^\dagger(\mathbf{p}')\} = \delta_{\mathbf{pp}'}\, \delta_{rr'} \tag{6.33}$$

$$\{a_r(\mathbf{p}), a_{r'}(\mathbf{p}')\} = \{a_r^\dagger(\mathbf{p}), a_{r'}^\dagger(\mathbf{p}')\} = 0 \tag{6.34}$$

$$\{b_r(\mathbf{p}), b_{r'}(\mathbf{p}')\} = \{b_r^\dagger(\mathbf{p}), b_{r'}^\dagger(\mathbf{p}')\} = 0 \tag{6.35}$$

$$\{a_r(\mathbf{p}), b_{r'}(\mathbf{p}')\} = \{a_r^\dagger(\mathbf{p}), b_{r'}^\dagger(\mathbf{p}')\} = 0 \tag{6.36}$$

$$\{a_r(\mathbf{p}), b_{r'}^\dagger(\mathbf{p}')\} = \{a_r^\dagger(\mathbf{p}), b_{r'}\mathbf{p}')\} \;\; = 0 \tag{6.37}$$

$$\{a_r^\dagger(\mathbf{p}), b_{r'}(\mathbf{p}')\} = \{a_r(\mathbf{p}), b_{r'}^\dagger\mathbf{p}')\} \;\; = 0 \tag{6.38}$$

Using Equation (6.33) in Equation (6.29) we obtain for the energy

$$E = \sum_{\mathbf{p}, r} \left[a_r^\dagger(\mathbf{p})\, a_r(\mathbf{p}) + b_r^\dagger(\mathbf{p})\, b_r(\mathbf{p}) - 1\right] E_p \tag{6.39}$$

$$= \sum_{\mathbf{p}, r} \left[N_r^{(1)}(\mathbf{p}) + N_r^{(2)}(\mathbf{p}) - 1\right] E_p \tag{6.40}$$

This result shows that energy is quantized where each energy quantum represents a particle or an anti-particle. As was the case for the electromagnetic field, also here there is a constant infinite term, except now that term is negative. We did not completely succeed in getting rid of negative energy as we set out to do, but if we ignore the infinitely large constant term, the other terms in Equation (6.40) have a sensible interpretation.

6.1.4 Momentum

We now calculate the momentum of a free Dirac particle described by the plane wave solutions Equation (6.2) and Equation (6.3). The momentum **P** is given by the expectation value of the operator $(\hbar/i)\nabla$

$$\mathbf{P} = \int d^3\mathbf{x}\, \psi^\dagger \mathbf{P} \psi \tag{6.41}$$

$$= \int d^3\mathbf{x}\, \overline{\psi}\, \frac{\gamma_4}{i} \nabla \psi \tag{6.42}$$

$$= \int \frac{d^3\mathbf{x}}{V} \sum_{\mathbf{p},\mathbf{p}',r,r'} \frac{m}{\sqrt{E_p E_p'}} \left[a_r^\dagger(\mathbf{p})\bar{u}_r(\mathbf{p})e^{-ipx} + b_r(\mathbf{p})\bar{v}_r(\mathbf{p})e^{ipx}\right] \frac{\gamma_4(i\mathbf{p}')}{i}$$
$$\left[a_{r'}(\mathbf{p}')u_{r'}(\mathbf{p}')e^{ip'x} - b_{r'}^\dagger(\mathbf{p}')v'(\mathbf{p}')e^{-ip'x}\right] \tag{6.43}$$

Following the same procedure as for the calculation of the energy we find

$$\mathbf{P} = \sum_{\mathbf{p},r} \frac{m}{E_p} \left[a_r^\dagger(\mathbf{p})\, a_r(\mathbf{p}) - b_r(\mathbf{p})\, b_r^\dagger(\mathbf{p})\right] \frac{E_p}{m} \mathbf{p} \tag{6.44}$$

$$= \sum_{\mathbf{p},r} \left[a_r^\dagger(\mathbf{p})\, a_r(\mathbf{p}) - b_r(\mathbf{p})\, b_r^\dagger(\mathbf{p})\right] \mathbf{p} \tag{6.45}$$

$$= \sum_{\mathbf{p},r} \left[a_r^\dagger(\mathbf{p})\, a_r(\mathbf{p}) + b_r^\dagger(\mathbf{p})\, b_r(\mathbf{p}) - 1\right] \mathbf{p} \tag{6.46}$$

$$= \sum_{\mathbf{p},r} \left[N_r^{(1)}(\mathbf{p}) + N_r^{(2)}(\mathbf{p}) - 1\right] \mathbf{p} \tag{6.47}$$

$$= \sum_{\mathbf{p},r} \left[N_r^{(1)}(\mathbf{p}) + N_r^{(2)}(\mathbf{p})\right] \mathbf{p} \tag{6.48}$$

where we have used the commutation relation from Equation (6.33) in Equation (6.45) and we have used Equation (6.30) and Equation (6.31) in Equation (6.46). The otherwise annoying -1 constant term has canceled in the summation over **p** because positive and negative values of **p** appear equally because we are quantizing in a box; compare the argument below Equation (1.67) for the electromagnetic field.

This result is equally remarkable as the result in Equation (6.40) for the same reasons.

6.1.5 Creation and Annihilation Operators

The creation and annihilation operators are postulated to satisfy the anti-commutation relations Equation (6.32) through Equation (6.38). They can be summarized as follows. All a, $a^\dagger$, b and $b^\dagger$ operators anticommute except the pairs $a_r(\mathbf{p}), a_r^\dagger(\mathbf{p})$ and $b_r(\mathbf{p}), b_r^\dagger(\mathbf{p})$ whose anticommutator equals 1. Contrary to the case of commutation relations in electromagnetism, we find now that the product of two of the same operators is zero when using Equation (6.32) through Equation (6.38)

$$a_r(\mathbf{p})\, a_r(\mathbf{p}) = a_r^\dagger(\mathbf{p})\, a_r^\dagger(\mathbf{p}) = b_r(\mathbf{p})\, b_r(\mathbf{p}) = b_r^\dagger(\mathbf{p})\, b_r^\dagger(\mathbf{p}) = 0 \tag{6.49}$$

We also find that (no summation over r is implied)

$$[N_r^{(1)}(\mathbf{p})]^2 = [a_r^\dagger(\mathbf{p})\, a_r(\mathbf{p})]\, [a_r^\dagger(\mathbf{p})\, a_r(\mathbf{p})] \tag{6.50}$$
$$= a_r^\dagger(\mathbf{p})[-a_r^\dagger(\mathbf{p})\, a_r(\mathbf{p}) + 1]a_r(\mathbf{p}) \tag{6.51}$$
$$= a_r^\dagger(\mathbf{p})\, a_r(\mathbf{p}) \tag{6.52}$$
$$= N_r^{(1)}(\mathbf{p}) \tag{6.53}$$

and similarly (no summation over r is implied)

$$[N_r^{(2)}(\mathbf{p})]^2 = N_r^{(2)}(\mathbf{p}) \tag{6.54}$$

These relations imply that the $N_r^{(1)}(\mathbf{p})$ and $N_r^{(2)}(\mathbf{p})$ take on the values 0 and 1 only. This is an expression of the Pauli exclusion principle for Dirac particles and anti-particles respectively.

So the requirement to have positive definite energy (with the caveat mentioned earlier) implies anticommutation relations for the a, $a^\dagger$, b and $b^\dagger$ operators, which in turn implies the Pauli exclusion principle: at most one particle with momentum $\mathbf{p}$ and spin coordinate r is allowed in the sums in Equation (6.40) and Equation (6.48).

We have not shown yet that the a, $a^\dagger$, b and $b^\dagger$ are indeed creation and annihilation operators. That this is so is not as automatic as it was in the case of the electromagnetic field where the $a_\lambda(\mathbf{k})$ and $a_\lambda^\dagger(\mathbf{k})$ satisfied the same commutation relations as the a and $a^\dagger$ did in the case of the Harmonic Oscillator, and we could appeal to the development of the Harmonic Oscillator formalism to show the properties of a and $a^\dagger$ in terms of annihilation and creation operators. In the present case $a, a^\dagger, b$ and $b^\dagger$ satisfy anticommutation relations. We will show, however, that $N_r^{(1)}(\mathbf{p})$ and $N_r^{(2)}(\mathbf{p})$ satisfy commutation relations with $a, a^\dagger$ and $b, b^\dagger$ respectively that are the same as those encountered in the case of the Harmonic Oscillator. We can then conclude that the a, b and $a^\dagger, b^\dagger$ are indeed annihilation and creation operators. We first consider the commutator

$$[b_r^\dagger(\mathbf{p})\, b_r(\mathbf{p}), a_{r'}(\mathbf{p}')] = b_r^\dagger(\mathbf{p})\, b_r(\mathbf{p})\, a_{r'}(\mathbf{p}') - a_{r'}(\mathbf{p}')\, b_r^\dagger(\mathbf{p})\, b_r(\mathbf{p}) \tag{6.55}$$
$$= b_r^\dagger(\mathbf{p})\, b_r(\mathbf{p})\, a_{r'}(\mathbf{p}') + b_r^\dagger(\mathbf{p})\, a_{r'}(\mathbf{p}')\, b_r(\mathbf{p}) \tag{6.56}$$
$$= b_r^\dagger(\mathbf{p})\, b_r(\mathbf{p})\, a_{r'}(\mathbf{p}') - b_r^\dagger(\mathbf{p})\, b_r(\mathbf{p})\, a_{r'}(\mathbf{p}') \tag{6.57}$$
$$= 0 \tag{6.58}$$

We moved $a_{r'}(\mathbf{p}')$ in the second term of Equation (6.55) in two steps to the right so that we cancel the first term. This implies that

$$[N_r^{(2)}(\mathbf{p}), a_{r'}(\mathbf{p}')] = 0 \tag{6.59}$$

Taking the Hermitian conjugate gives

$$[N_r^{(2)}(\mathbf{p}), a_{r'}^{\dagger}(\mathbf{p}')] = 0 \tag{6.60}$$

Similarly we can show that

$$[a_r^{\dagger}(\mathbf{p})\, a_r(\mathbf{p}), b_{r'}(\mathbf{p}')] = 0 \tag{6.61}$$

and this implies that

$$[N_r^{(1)}(\mathbf{p}), b_{r'}(\mathbf{p}')] = 0 \tag{6.62}$$

Taking the Hermitian conjugate gives

$$[N_r^{(1)}(\mathbf{p}), b_{r'}^{\dagger}(\mathbf{p}')] = 0 \tag{6.63}$$

Next consider the commutator $[N_r^{(1)}(\mathbf{p}), a_{r'}(\mathbf{p}')]$. We find

$$[N_r^{(1)}(\mathbf{p}), a_{r'}(\mathbf{p}')] = [a_r^{\dagger}(\mathbf{p})a_r(\mathbf{p}), a_{r'}(\mathbf{p}')] \tag{6.64}$$

$$= a_r^{\dagger}(\mathbf{p})\, a_r(\mathbf{p})\, a_{r'}(\mathbf{p}') - a_{r'}(\mathbf{p}')\, a_r^{\dagger}(\mathbf{p})\, a_r(\mathbf{p}) \tag{6.65}$$

$$= a_r^{\dagger}(\mathbf{p})\, a_r(\mathbf{p})\, a_{r'}(\mathbf{p}') - (-a_r^{\dagger}(\mathbf{p})\, a_{r'}(\mathbf{p}') + \delta_{\mathbf{p}'\mathbf{p}}\delta_{r'r})\, a_r(\mathbf{p}) \tag{6.66}$$

$$= a_r^{\dagger}(\mathbf{p})\, a_r(\mathbf{p})\, a_{r'}(\mathbf{p}') + a_r^{\dagger}(\mathbf{p})\, a_{r'}(\mathbf{p}')\, a_r(\mathbf{p}) - \delta_{\mathbf{p}'\mathbf{p}}\, \delta_{r'r}\, a_r(\mathbf{p}) \tag{6.67}$$

$$= a_r^{\dagger}(\mathbf{p})\, a_r(\mathbf{p})\, a_{r'}(\mathbf{p}') - a_r^{\dagger}(\mathbf{p})\, a_r(\mathbf{p})\, a_{r'}(\mathbf{p}') - \delta_{\mathbf{p}'\mathbf{p}}\, \delta_{r'r}\, a_r(\mathbf{p}) \tag{6.68}$$

$$= -\delta_{\mathbf{p}\mathbf{p}'}\, \delta_{rr'}\, a_r(\mathbf{p}) \tag{6.69}$$

In the second term in Equation (6.65) we moved $a_{r'}(\mathbf{p}')$ in two steps to the right so that we cancel the first term. Taking the Hermitian conjugate gives

$$[N_r^{(1)}(\mathbf{p}), a_{r'}^{\dagger}(\mathbf{p}')] = \delta_{\mathbf{p}\mathbf{p}'}\, \delta_{rr'}\, a_r^{\dagger}(\mathbf{p}) \tag{6.70}$$

Note the change of sign on the right-hand side.

A similar result holds for $[N_r^{(2)}(\mathbf{p}), b_{r'}(\mathbf{p}')]$:

$$[N_r^{(2)}(\mathbf{p}), b_{r'}(\mathbf{p}')] = [b_r^{\dagger}(\mathbf{p})\, b_r(\mathbf{p}), b_{r'}(\mathbf{p}')] \tag{6.71}$$

$$= -\delta_{\mathbf{p}\mathbf{p}'}\, \delta_{rr'}\, a_r(\mathbf{p}) \tag{6.72}$$

Taking the Hermitian conjugate gives

$$[N_r^{(2)}(\mathbf{p}), b_{r'}^{\dagger}(\mathbf{p}')] = \delta_{\mathbf{p}\mathbf{p}'}\, \delta_{rr'}\, b_r^{\dagger}(\mathbf{p}) \tag{6.73}$$

Using Equation (6.40), and replacing E by H, and Equation (6.69) and Equation (6.59) we find

$$[H, a_r(\mathbf{p})] = -a_r(\mathbf{p})E_p \tag{6.74}$$

Taking the Hermitian conjugate gives

$$[H, a_r^\dagger(\mathbf{p})] = a_r^\dagger(\mathbf{p})E_p \tag{6.75}$$

Similarly, using Equation (6.40), and replacing E by H, and Equation (6.72) and Equation (6.62) we find

$$[H, b_r(\mathbf{p})] = -b_r(\mathbf{p})E_p \tag{6.76}$$

Taking the Hermitian conjugate gives

$$[H, b_r^\dagger(\mathbf{p})] = b_r^\dagger(\mathbf{p})E_p \tag{6.77}$$

Even though the $a, a^\dagger, b$ and $b^\dagger$ satisfy anticommutation relations, we find the same commutation relations between $N^{(1)}, N^{(2)}, H$ and a, $a^\dagger$, b and $b^\dagger$ that apply to the Harmonic Oscillator. So once again we can take over the results of the Harmonic Oscillator and consider the $a, b, a^\dagger$ and $b^\dagger$ annihilation and creation operators decreasing and increasing the number of quanta (particles or anti-particles). Thus a positron with momentum $\mathbf{p}$ and spin label r is represented by

$$|\mathbf{p}, r\rangle = b_r^\dagger(\mathbf{p})|0\rangle \tag{6.78}$$

and two electrons with momentum and spin labels $\mathbf{p}_1, r_1$ and $\mathbf{p}_2, r_2$ respectively are represented by

$$|\mathbf{p}_1, r_1; \mathbf{p}_2, r_2\rangle = a_{r_1}^\dagger(\mathbf{p}_1)a_{r_2}^\dagger(\mathbf{p}_2)|0\rangle \tag{6.79}$$

We see from Equation (6.79) that permuting the labels 1 and 2 gives a minus sign, that is the state Equation (6.79) is antisymmetric, as it should be for two Fermions. Setting $\mathbf{p}_1 = \mathbf{p}_2$ and $r_1 = r_2$ in Equation (6.79), we find that $|\mathbf{p}_1, r_1; \mathbf{p}_2, r_2\rangle = 0$ as it should be according to the Pauli exclusion principle.

We are now in a position to start the study of Quantum Electrodynamics but before we do that we need to discuss second quantization of the massive spin 1 field. This plays the role of mediator of the charged weak and neutral weak interaction in the same way as the photon is the mediator of the electromagnetic interaction.

6.2 SECOND QUANTIZATION OF SPIN 1 FIELDS

The W-boson mediates the charged weak interaction while the Z-boson mediates the neutral weak interaction. The W-boson is charged and the Z-boson is neutral like the photon. Both have spin 1 like the photon, but they are massive($m_W = 80.4\,\text{GeV}$ and $m_Z = 91.2\,\text{GeV}$), unlike the photon. To accommodate the spin of the W and Z we must allow for internal degrees of freedom like those of the electromagnetic vector potential A_μ. We therefore write W_μ and Z_μ for the fields that correspond to the W and Z bosons. Each component of W_μ and Z_μ satisfies the Einstein relation Equation (4.1) and therefore also Equation (4.4) so we have

$$(\partial^2 - m_W^2)\, W_\mu(x) = 0 \qquad (\partial^2 - m_Z^2)\, Z_\mu(x) = 0 \tag{6.80}$$

The index μ labels the internal degree of freedom of the W and Z fields (the spin of the W and Z); compare with the electromagnetic vector field A_μ. In the case of theelectromagnetic field we exploited the invariance of **E** and **B** under Gauge transformations of A_μ to impose the Lorenz condition in Equation (1.12) $\partial_\mu A_\mu = 0$. The Lorenz condition relates the four components of A_μ so that only three independent components remain, a situation suitable for the description of a spin 1 particle. We made a further Gauge transformation to the Coulomb Gauge to show that there are only two and not three independent components. It is seen that invariance under Gauge transformations depends upon the fact that the electromagnetic field A_μ is massless (and thus describes massless photons). In the present case of the W and Z fields we impose the Lorenz condition with the motivation that also here we want W_μ and Z_μ to have three independent components, not four, to describe massive spin 1 particles. Thus we require

$$\partial_\mu W_\mu = 0 \qquad \partial_\mu Z_\mu = 0 \tag{6.81}$$

To see how this works, substitute a plane wave solution of the form

$$W_\mu(x) = \sum_{r=1}^{4} a_r(\mathbf{p})[\varepsilon_r(\mathbf{p})]_\mu e^{ipx} \tag{6.82}$$

in Equations (6.80) and (6.81). Here r labels the four basis vectors $\varepsilon_r(\mathbf{p})$ each of whose components are labeled by μ. In the first case we recover the Einstein relation that was input to Equation (6.80), while in the second case we get

$$\sum_{r=1}^{4} a_r(\mathbf{p})[\varepsilon_r(\mathbf{p})]_\mu p_\mu = 0 \tag{6.83}$$

In the centre-of-mass of the W or the Z we have $p_\mu = (0, im)$ with m either the W or Z mass. In the center-of-mass we can choose four and only four linearly independent unit four-vectors ε_r, $r = 1, 4$

$$(\varepsilon_1)_\mu = (1,0,0,0) \quad (\varepsilon_2)_\mu = (0,1,0,0) \quad (\varepsilon_3)_\mu = (0,0,1,0) \quad (\varepsilon_4)_\mu = (0,0,0,1) \tag{6.84}$$

corresponding to four polarization directions. Using $p_\mu = (0, im)$ with $m \neq 0$ we find from Equation (6.83) that $a_4 = 0$. This shows that indeed only three components of a_μ and W_μ and Z_μ are non-zero in the centre-of-mass. That there are three polarizations and not four should be a Lorenz invariant notion. Thus we expect that also in the laboratory frame three linearly independent polarizations will suffice. Of course the Lorenz conditions Equation (6.81), being Lorenz invariant, are valid in the laboratory. So for a W or Z moving in the z-direction with $p_\mu = (0, 0, p, iE)$ we define three linearly independent unit four-vectors for the three polarizations as

$$(\varepsilon_1)_\mu = (1,0,0,0) \qquad (\varepsilon_2)_\mu = (0,1,0,0) \qquad (\varepsilon_3)_\mu = (0,0,a,ib) \tag{6.85}$$

We have chosen ε_1 and ε_2 to be perpendicular to the three-momentum of the W or Z and perpendicular to ε_3 and we have given them unit length. We choose a and b such that the Lorenz condition in Equation (6.81) is satisfied and that ε_3 has unit length. We obtain $ap - bE = 0$ and $a^2 - b^2 = 1$. Taking the positive solution (the overall minus sign for the other solution does not correspond to a different solution) we find $a = E/m$ and $b = p/m$ and Equation (6.85) becomes

$$(\varepsilon_1)_\mu = (1,0,0,0) \quad (\varepsilon_2)_\mu = (0,1,0,0) \quad (\varepsilon_3)_\mu = \left(0, 0, \frac{E}{m}, i\frac{p}{m}\right) \tag{6.86}$$

It is seen that Equation (6.86) reduces to Equation (6.84) in the center-of-mass system as expected. The relations Equation (6.86) cannot be obtained from the relations Equation (6.85) by a Lorenz transformation. This is an indication that there are deeper issues behind the arguments used above to reduce the four independent components to three. If we apply the present reasoning to the photon field we see that Gauge invariance corresponds to the separation of the photon field in a spin 0 and a spin 1 field. We note for future use that the polarization vectors are orthogonal to each other and that they satisfy the relation

$$\sum_{r=1}^{3} [\varepsilon_r(\mathbf{p})]_\mu [\varepsilon_r^*(\mathbf{p})]_\nu = \delta_{\mu\nu} + \frac{p_\mu p_\nu}{m^2} \tag{6.87}$$

This relation can be derived as follows. The left-hand side must be a 4×4 matrix with indices μ and ν that is symmetric in μ and ν. The only quantities available to construct such a matrix are the Kronecker $\delta_{\mu\nu}$ and $p_\mu p_\nu$. Therefore we set

$$\sum_{r=1}^{3}[\varepsilon_r(\mathbf{p})]_\mu[\varepsilon_r^*(\mathbf{p})]_\nu = a\delta_{\mu\nu} + bp_\mu p_\nu \tag{6.88}$$

where a and b are to be determined. To determine them we set $\mu = \nu$ and sum over $\mu = \nu$ and r using Equation (6.86). We obtain the relation $3 = 4a + bp^2$. Note that complex conjugation here does *not* imply that we take the complex conjugate of the fourth component of $(\varepsilon_3)_\mu$, see the remark below. Next we multiply Equation (6.88) by p_μ and sum over μ, using Equation (6.86) for the polarization vectors. It follows that $0 = p_\nu(a + bp^2)$. Solving for a and b from these two conditions, we find $a = 1$ and $b = -1/p^2 = 1/m^2$. We note an important property of Equation (6.87) when it is evaluated in the W or Z center-of-mass. Using that in the center-of-mass $p_\mu = (0, im)$ we find that the right-hand side of Equation (6.87) equals zero when either $\mu = 4$ or $\nu = 4$ or when $\mu = \nu = 4$.

As mentioned above, the complex conjugate of $[\varepsilon_t(\mathbf{p})]_\mu$ is equal to itself. This is one of the rare cases where our metric leads to inconsistent use. The complex character of polarization vectors as shown by the * in Equation (6.87) and Equation (6.88) appears when we are considering circularly polarized W or Z (which we are not in the present application). A discussion of this is found in the section about polarization and spin near Equation (1.68) and Equation (1.69). Thus complex conjugation is not to be applied to the imaginary fourth component of a four-vector.

Because the equations in Equation (6.80) are linear and homogeneous, the most general solutions to Equation (6.80) can be written as a linear superposition of plane waves

$$W_\mu(x) = \frac{1}{\sqrt{V}}\sum_{\mathbf{p},r}\frac{1}{\sqrt{2E_p}}\left[a_r(\mathbf{p})\,e^{ipx}[\varepsilon_r(\mathbf{p})]_\mu + b_r^\dagger(\mathbf{p})\,e^{-ipx}[\varepsilon_r^*(\mathbf{p})]_\mu\right] \tag{6.89}$$

$$W_\mu^\dagger(x) = \frac{1}{\sqrt{V}}\sum_{\mathbf{p},r}\frac{1}{\sqrt{2E_p}}\left[a_r^\dagger(\mathbf{p})\,e^{-ipx}[\varepsilon_r^*(\mathbf{p})]_\mu + b_r(\mathbf{p})\,e^{ipx}[\varepsilon_r(\mathbf{p})]_\mu\right] \tag{6.90}$$

and

$$Z_\mu(x) = \frac{1}{\sqrt{V}}\sum_{\mathbf{p},r}\frac{1}{\sqrt{2E_p}}\left[a_r(\mathbf{p})\,e^{ipx}[\varepsilon_r(\mathbf{p})]_\mu + a_r^\dagger(\mathbf{p})\,e^{-ipx}[\varepsilon_r^*(\mathbf{p})]_\mu\right] \tag{6.91}$$

with $E_p = \sqrt{\mathbf{p}^2 + m^2}$, compare with Equation (1.30) and Equations (6.2) and (6.3) for the electromagnetic and Dirac fields. The $\varepsilon_r(\mathbf{p})$ have been discussed earlier and are given by Equation (6.86). As in the case of the photon field, we indicate the complex conjugate of ε in the second term in Equation (6.89), Equation (6.90) and Equation (6.91) to allow for circularly polarized solutions of Equations (6.80). The summation over r is from 1 to 3, not 1 to 2 or 1 to 4; compare with the case of the electromagnetic field. The prefactor $1/\sqrt{V}$ and the factors $1/\sqrt{2E_p}$ merely redefine the operators $a_r(\mathbf{p})$ and $b_r^\dagger(\mathbf{p})$. They are chosen such that the energy and momentum of the W and Z fields are sensible.

By writing Hermitian conjugates instead of complex conjugates in Equation (6.89), Equation (6.90) and Equation (6.91) we anticipate that the a and b will be operators. Because the W and Z are bosons, we postulate, as we did before for the electromagnetic field, that $a, b, a^\dagger$ and $b^\dagger$ satisfy *commutation* relations (and not anti-commutation relations)

$$[a_r(\mathbf{p}), a_{r'}^\dagger(\mathbf{p}')] = \delta_{\mathbf{pp}'}\,\delta_{rr'} \tag{6.92}$$

$$[b_r(\mathbf{p}), b_{r'}^\dagger(\mathbf{p}')] = \delta_{\mathbf{pp}'}\,\delta_{rr'} \tag{6.93}$$

$$[a_r(\mathbf{p}), a_{r'}(\mathbf{p}')] = [a_r^\dagger(\mathbf{p}), a_{r'}^\dagger(\mathbf{p}')] = 0 \tag{6.94}$$

$$[b_r(\mathbf{p}), b_{r'}(\mathbf{p}')] = [b_r^\dagger(\mathbf{p}), b_{r'}^\dagger(\mathbf{p}')] = 0 \tag{6.95}$$

$$[a_r(\mathbf{p}), b_{r'}(\mathbf{p}')] = [a_r^\dagger(\mathbf{p}), b_{r'}^\dagger(\mathbf{p}')] = 0 \tag{6.96}$$

$$[a_r(\mathbf{p}), b_{r'}^\dagger(\mathbf{p}')] = [a_r^\dagger(\mathbf{p}), b_{r'}\mathbf{p}')] \;= 0 \tag{6.97}$$

$$[a_r^\dagger(\mathbf{p}), b_{r'}(\mathbf{p}')] = [a_r(\mathbf{p}), b_{r'}^\dagger\mathbf{p}')] \;= 0 \tag{6.98}$$

Anti-commutation relations in analogy with Equation (6.32) through Equation (6.38) would have imposed the Pauli exclusion principle on the W and Z bosons, but the latter are not subject to the Paul exclusion principle. We arbitrarily label the W^- a particle and the W^+ an anti-particle. Therefore, $a_r^\dagger(\mathbf{p})$ and $b_r^\dagger(\mathbf{p})$ create a W^- and a W^+ respectively while $a_r(\mathbf{p})$ and $b_r(\mathbf{p})$ annihilate a W^- and a W^+ respectively, with momentum $\mathbf{p}$ and polarization index r. Thus the field $W_\mu(x)$ in Equation (6.89) increases the electric charge by one unit, either by annihilation of a W^- boson or by creation of a W^+ anti-particle. Likewise, the field $W_\mu^\dagger(x)$ decreases the electric charge by one unit. Compare with the Dirac fields Equation (6.2) and its Hermitian conjugate Equation (6.3). The structure of $Z_\mu(x)$ in Equation (6.91) is the same as for the photon field $A_\mu(x)$, namely it can annihilate or create Z^0 particles, conserving charge in the process. The expressions for $A_\mu(x)$ and $Z_\mu(x)$ are Hermitian while the expressions for $\psi(x)$ and W_μ are not. The relationship between Hermiticity and charge can be inferred from Equation (4.10) using the fact that the electric current density is proportional to the probability current density which is seen to vanish for real fields.

We will not repeat in detail the discussion of one and more particle states because it closely follows the discussion in the study of the Harmonic

Oscillator and the electromagnetic field. The occupation number operators can be shown to take on the value zero or any positive value as expected for bosons. Multiparticle states can be constructed by sequential application of creation operators to the vacuum.

PROBLEMS

(1) Derive Equations (6.5) through (6.8).
(2) Derive Equations (6.13) through (6.16).

7

Covariant Perturbation Theory and Applications

7.1 COVARIANT PERTURBATION THEORY

7.1.1 Hamiltonian Density

We have developed time-dependent perturbation theory in Section 1.3. The results were clearly not relativistically correct because the time variable played a role different from the space variables. In particular, integrations over time appeared without similar integrations over space. In relativistic quantum field theory we will deal with interactions whose Hamiltonian is described by a Hamiltonian density $\mathcal{H}$. The Hamiltonian density is defined by the relation

$$H = \int d^3\mathbf{x}\,\mathcal{H} \tag{7.1}$$

where H is the Hamiltonian and $\mathcal{H}$ the Hamiltonian density, an operator like H. The integrations over time that we encountered earlier will now be complemented by an integration over space giving an integration over d^4x, treating space and time coordinates on an equal footing.

We will give a number of examples of Hamiltonian densities taken from the electromagnetic interaction (mediated by the electromagnetic field A) and the weak interaction. Of the latter we will give two examples: the Hamiltonian density of the charged weak interaction (mediated by the W^+ and W^-) and the neutral weak interaction (mediated by the Z^0). These four bosons all have spin 1. We will consider how these four bosons interact with leptons and quarks.

An Introduction to Advanced Quantum Physics Hans P. Paar

The interactions can be represented by an interaction Hamiltonian density that has the form of the four-vector product of a current density $j_\mu(x)$ and one of the boson fields A_μ, $W_\mu^\pm$ or Z_μ^0 and is thus a scalar. The current density involves lepton and quark fields $\overline{\psi}$ and ψ. Leptons and quarks are fermions with spin $\frac{1}{2}$ and are the quanta of the Dirac field. A lepton is a particle that is not subject to the strong interaction, while a quark *is* subject to it. The lepton and quark flavors each come in three doublets: (e^-, ν_e), (μ^-, ν_μ) and (τ^-, ν_τ) for the lepton sector, and (u, d), (c, s) and (t, b) for the quark sector. The first lepton doublet consists of the electron and the electron-neutrino, the second lepton doublet consists of the muon and the muon-neutrino, and the third lepton doublet consists of the tau and tau-neutrino. The first quark doublet consists of the up and the down quark, the second quark doublet consists of the charm and the strange quark, and the third quark doublet consists of the top and the bottom quark. The charged leptons have one unit of negative charge as shown, while the neutrinos are neutral. The u, c and t quarks have charge $\frac{2}{3}|e|$ while the d, s and b quarks have charge $-\frac{1}{3}|e|$. For each doublet of particles there exists a corresponding doublet of anti-particles. These are also spin $\frac{1}{2}$ of course, and the charged members have opposite charge. Different particle species are said to have different flavors. A discussion of these fermions and anti-fermions is outside the scope of this book.

The Hamiltonian density for the electromagnetic interaction was found at the time the Dirac equation was invented. The Hamiltonian densities for the charged and neutral weak interactions were determined after an extensive experimental program over the period 1950–1990. A discussion of this matter is also outside the scope of this book and we will simply state the expressions for their respective Hamiltonian densities.

These expressions have been constructed in such a manner that the experimentally verified conservation laws of charge, fermion number, baryon number, and lepton number are satisfied. A discussion of these quantum numbers is also outside the scope of this book. In addition the experimentally verified conservation of particle flavor (species) in the electromagnetic interaction and the neutral weak interaction are built in their respective Hamiltonian densities. It is found experimentally that the charged weak interaction changes particle flavor in a precisely prescribed manner and its rules are also built in its respective Hamiltonian density. We now discuss the Hamiltonian densities of the electromagnetic, charged weak and neutral weak interactions. The strong interaction is outside the scope of this book (as is the gravitational interaction).

The Hamiltonian density of the electromagnetic interaction is given by

$$\mathcal{H}^{\text{em}} = -j_\mu^{\text{em}} A_\mu \tag{7.2}$$

where the electromagnetic current j_μ^{em} is given by

$$j_\mu^{\text{em}} = iq\overline{\psi}\gamma_\mu\psi \tag{7.3}$$

with q the charge, see Equation (4.72), but here called a coupling constant in analogy with coupling constants to be introduced below. A_μ represents the photon, the mediator of the electromagnetic interaction. The $\overline{\psi}$ and ψ refer to the same particle flavor, so that the electromagnetic interaction obeys the experimentally established conservation laws listed above. Because charge is conserved, the photon is neutral. The Hamiltonian density Equation (7.2) describes the interaction of the electromagnetic current j_μ^{em} and the electromagnetic field A_μ. The electromagnetic current is conserved, that is $\partial_\mu j_\mu^{\text{em}} = 0$.

There is an electromagnetic current for each charged lepton flavor ℓ and quark flavor q so we can write in general

$$j_\mu^{\text{em}} = \sum_{i=\ell,q} (j_\mu^{\text{em}})_i \tag{7.4}$$

$$(j_\mu^{\text{em}})_i = i\,q_i\overline{\psi}_i\gamma_\mu\psi_i \tag{7.5}$$

where q_i is the charge of the particle with flavor i. Therefore, the neutrinos have no electromagnetic interaction and they do not appear in the sum in Equation (7.4). $\mathcal{H}^{\text{em}}$ is seen to be Hermitian and the q_i are dimensionless coupling constants.

The Hamiltonian density of the charged weak interaction is given by

$$\mathcal{H}^{\text{wk,ch}} = j_\mu^{\text{wk,ch}} W_\mu^\dagger + (j_\mu^{\text{wk,ch}})^\dagger W_\mu \tag{7.6}$$

where the charged weak current $j_\mu^{\text{wk,ch}}$ has the general structure

$$j_\mu^{\text{wk,ch}} = \overline{\psi}(G_V^{\text{wk,ch}}\gamma_\mu + g_A^{\text{wk,ch}}\gamma_\mu\gamma_5)\psi \tag{7.7}$$

with $\gamma_5 = \gamma_1\gamma_2\gamma_3\gamma_4$. W_μ represents the W boson, the mediator of the charged weak interaction. The g_V and g_A are dimensionless coupling constants and have no analogue in classical physics, contrary to the electromagnetic interaction. The $\overline{\psi}$ and ψ refer to two different particles from the same doublet, so that the experimentally established conservation laws of the charged weak current listed above are obeyed. Particles in the same doublet differ in charge by one unit, hence the name *charged* weak current. The lepton doublets are (e^-, ν_ℓ), (μ^-, ν_μ), (τ^-, ν_τ) and similar doublets for their anti-particles. The quark doublets for the purpose of the charged weak interaction are (u, d_R), (c, s_R), (t, b_R) and similar doublets for their anti-particles. The d_R, s_R, b_R are related to the d, s, b by a unitary three-dimensional matrix called the Cabibbo-Kobayashi-Maskawa or CKM matrix after the physicists who recognized the need for it. Thus the kets representing d_R, s_R, b_R respectively are each linear superpositions of the kets representing d, s, b quarks. The coefficients are elements of the CKM matrix and are determined experimentally. The index R stands for 'Rotated' in recognition of the fact that a unitary matrix with real elements represents a rotation. The CKM matrix, however, has some complex elements, a property that allows for a particle–anti-particle asymmetry that has been observed experimentally and is part of the explanation for the matter–antimatter asymmetry observed in the Universe. The Hamiltonian density is constructed so that it changes a particle's flavor in a precisely prescribed manner referred to above. Because charge is conserved, the W particle carries one unit of positive or negative charge.

It is easy to verify that γ_5 is given by

$$\gamma_5 = \begin{pmatrix} 0 & -1 \\ -1 & 0 \end{pmatrix} \tag{7.8}$$

and thus that γ_5 is Hermitian. One can verify that

$$\{\gamma_\mu, \gamma_5\} = 0 \qquad \mu = 1, 4 \tag{7.9}$$

so all five gamma matrices satisfy

$$\{\gamma_\mu, \gamma_\nu\} = \delta_{\mu\nu} \qquad \mu, \nu = 1, 5 \tag{7.10}$$

The Hamiltonian density Equation (7.6) describes the interaction of the charged weak current and the W boson and conserves charge, lepton number and baryon number.

There is a charged weak current for each lepton doublet L and quark doublet Q so we write instead of Equation (7.7)

$$j_\mu^{\text{wk,ch}} = \sum_{i=L,Q} (j_\mu^{\text{wk,ch}})_i \tag{7.11}$$

$$(j_\mu^{\text{wk,ch}})_i = \overline{\psi}_i[(g_V^{\text{wk,ch}})_i\gamma_\mu + (g_A^{\text{wk,ch}})_i\gamma_\mu\gamma_5]\psi_i \tag{7.12}$$

Note that $\mathcal{H}^{\text{wk,ch}}$ is Hermitian even though $j_\mu^{\text{wk,ch}}$ and W_μ are not.

The Hamiltonian density of the neutral weak interaction is given by

$$\mathcal{H}^{\text{wk,n}} = j_\mu^{\text{wk,n}} Z_\mu \tag{7.13}$$

where the neutral weak current $j_\mu^{\text{wk,n}}$ has the general structure

$$j_\mu^{\text{wk,n}} = \overline{\psi}(g_V^{\text{wk,n}} \gamma_\mu + g_A^{\text{wk,n}} \gamma_\mu \gamma_5)\psi \tag{7.14}$$

and Z_μ represents the Z boson, the mediator of the neutral weak interaction. The g_V and g_A are dimensionless coupling constants and have no analogue in classical physics, contrary to the electromagnetic interaction. The $\overline{\psi}$ and ψ refer to the same particle flavor so that the neutral weak interaction obeys the experimentally established conservation laws listed above. As was the case with the electromagnetic interaction, the neutral weak interaction cannot change the flavor of the particles, compare with the electromagnetic current Equation (7.3) and the charged weak current Equation (7.7). Because charge is conserved the Z is a neutral particle, and hence the name *neutral (weak) current* for the current in Equation (7.14). The Hamiltonian density Equation (7.13) describes the interaction of the neutral weak current and the Z boson, and by construction conserves particle flavor and thus conserves charge, lepton number and baryon number.

There is a neutral weak current for each lepton flavor ℓ and quark flavor q so we can write in general

$$j_\mu^{\text{wk,n}} = \sum_{i=\ell,q} (j_\mu^{\text{wk,n}})_i \tag{7.15}$$

$$(j_\mu^{\text{wk,n}})_i = i\,\overline{\psi}_i(g_V^{\text{wk,n}} \gamma_\mu + g_A^{\text{wk,n}} \gamma_\mu \gamma_5)\psi_i \tag{7.16}$$

Note that $\mathcal{H}^{\text{wk}}$ is Hermitian. Also here, the $g_V^{\text{wk,n}}$ and $g_A^{\text{wk,n}}$ are dimensionless coupling constants and have no analogue in classical physics, contrary to the electromagnetic interaction. Note the similarity (and dissimilarity) between the form of the electromagnetic and the two weak interaction Hamiltonian densities Equation (7.2), Equation (7.6) and Equation (7.13), and between the electromagnetic current Equation (7.3) and the neutral weak currents Equation (7.7) and Equation (7.14). It is remarkable that the electromagnetic and the two weak Hamiltonian densities are all a product of a current and a vector boson field. The electromagnetic current and the weak currents differ from each other by the term with the γ_5, typical of the weak interaction. It can be shown that the simultaneous appearance of terms with γ_μ and $\gamma_\mu\gamma_5$ leads to parity violation. The electromagnetic and neutral weak currents have been unified in the Standard Model of elementary particle physics.

7.1.2 Interaction Representation

Thus far, state vectors were time dependent while operators were in most, but not all, cases time independent. This is called the Schrödinger representation (SR). The time dependence of the state vector $|\psi(t)\rangle$ was specified by the Schrödinger equation

$$-\frac{1}{i}\frac{\partial|\psi(t)\rangle^{SR}}{\partial t} = H|\psi(t)\rangle^{SR} \tag{7.17}$$

which has the formal solution

$$|\psi(t)\rangle^{SR} = e^{-iHt}|\psi(0)\rangle^{SR} \tag{7.18}$$

We consider a Hamiltonian which is the sum of a 'large' time-independent piece H_0 and a 'small' time-dependent piece $H_1(t)$ where $H_1(t)$ describes the interaction under consideration. In the absence of an interaction ($H_1(t) = 0$) the state vector's time development is

$$|\psi(t)\rangle^{SR} = e^{-iH_0t}|\psi(0)\rangle^{SR} = U(t)|\psi(0)\rangle^{SR} \tag{7.19}$$

where

$$U(t) = e^{-iH_0t} \tag{7.20}$$

Note that the operator $U(t)$ is unitary, so $UU^\dagger = U^\dagger U = 1$, because H_0 is Hermitian.

The interaction representation (IR) is obtained by transforming state vectors and operators as

$$|\psi(t)\rangle^{IR} = U^\dagger(t)|\psi(t)\rangle^{SR} \tag{7.21}$$

$$\mathcal{O}(t)^{IR} = U^\dagger(t)\,\mathcal{O}^{SR}U(t) \tag{7.22}$$

with U given by Equation (7.20). The motivation for the definition in Equation (7.21) is to make the state vector in the IR time independent if there is no interaction. Substituting Equation (7.19) in Equation (7.21) we get

$$|\psi(t)\rangle^{IR} = |\psi(0)\rangle^{SR} \tag{7.23}$$

which shows that indeed the state vector in the IR is independent of time when $H_1(t) = 0$. Another motivation for the definitions Equation (7.21) and Equation (7.22) is to make relations between operators and between operators and state vectors form-independent or form-invariant. For example, if the relation $C^{SR} = A^{SR}B^{SR}$ holds in the SR for operators A, B, C, it also holds in the IR. If we multiply the equality by $U^\dagger$ from the left and U from the right and insert $UU^\dagger(= 1)$ between A^{SR} and B^{SR}, we get $U^\dagger C^{SR}U = U^\dagger A^{SR}UU^\dagger B^{SR}U$. Using Equation (7.22) we see that the relation also holds in the IR. A similar argument shows that if $C^{SR} = A^{SR} + B^{SR}$ then also $C^{IR} = A^{IR} + B^{IR}$. Mixed expressions are also form invariant. If $|\psi\rangle^{SR} = \mathcal{O}^{SR}|\phi\rangle^{SR}$ and we multiply this equality from the left by $U^\dagger$, we get $U^\dagger|\psi\rangle^{SR} = U^\dagger\mathcal{O}^{SR}|\phi\rangle^{SR}$. Inserting $UU^\dagger(= 1)$ between $\mathcal{O}^{SR}$ and $|\phi\rangle^{SR}$

we get $U^\dagger|\psi\rangle^{\text{SR}} = U^\dagger \mathcal{O}^{\text{SR}} U U^\dagger |\phi\rangle^{\text{SR}}$ or $|\psi\rangle^{\text{IR}} = \mathcal{O}^{\text{IR}}|\phi\rangle^{\text{IR}}$ where we used Equation (7.22). The transformations Equation (7.21) and Equation (7.22) are called similarity transformations. Of course the factors that appear in the relations in say the IR are entirely different from the factors that appear in the 'similar' relation in the SR. Hence the usage of the term 'form independent' above.

We now cast the Schrödinger Equation (7.17), with the full Hamiltonian $H = H_0 + H_1(t)$, in a form that involves the state vector in the IR. To this end we invert Equation (7.21) and substitute the resulting $|\psi(t)\rangle^{\text{SR}} = U(t)|\psi(t)\rangle^{\text{IR}}$ in Equation (7.17). This gives

$$-\frac{1}{i}\frac{dU}{dt}|\psi(t)\rangle^{\text{IR}} - \frac{1}{i}U\frac{\partial|\psi(t)\rangle^{\text{IR}}}{\partial t} = HU|\psi(t)\rangle^{\text{IR}} \tag{7.24}$$

Calculating the derivative of $U(t)$ using Equation (7.20) and writing $H = H_0 + H_1(t)$, two terms proportional to H_0 cancel and we get

$$-\frac{1}{i}U\frac{\partial|\psi(t)\rangle^{\text{IR}}}{\partial t} = H_1 U|\psi(t)\rangle^{\text{IR}} \tag{7.25}$$

If we multiply Equation (7.25) by $U^\dagger$ from the left we get

$$-\frac{1}{i}\frac{\partial|\psi(t)\rangle^{\text{IR}}}{\partial t} = H_1^{\text{IR}}|\psi(t)\rangle^{\text{IR}} \tag{7.26}$$

with

$$H_1^{\text{IR}} = U^\dagger H_1 U \tag{7.27}$$

where we used Equation (7.22). We see that if $H_1(t) = 0$ (a non-interacting system) then $|\psi(t)\rangle^{\text{IR}}$ is independent of time, as expected for a state vector in the IR.

Henceforth we will work in the IR and we do not write the 'IR' label. Working in the IR has the advantage that the time dependence of state vectors is entirely due to the interaction described by $H_1(t)$. The interaction Hamiltonian is form-invariant, so if we know its form in the SR, we also know it in the IR, provided that we replace the operators in their SR form by their equivalent form in the IR. Because relations between operators are form-invariant, commutation relations are the same in the SR and the IR. Therefore, creation and annihilation operators satisfy the usual commutation relations valid for non-interacting fields, and the formalism we developed in the previous chapters for free fields remain valid in the IR. In particular, solutions of the wave equation in terms of plane waves, valid for non-interacting fields, remain valid in the IR.

7.1.3 Covariant Perturbation Theory

We will now find the solution of Equation (7.26) with the initial condition

$$|\psi(t)\rangle = |\psi(t_0)\rangle \quad \text{at} \quad t = t_0 \tag{7.28}$$

This differential equation can usually only be solved iteratively. To this end we consider $H_1(t)$ a perturbation and write it as $\lambda H_1(t)$ where the 'small' parameter λ is used to keep track of the order of various terms. There is a close analogy with the time dependent perturbation theory in Section 1.3. As we did there, we expand the state vector in powers of the perturbation as

$$|\psi(t)\rangle = |\psi^{(0)}(t)\rangle + \lambda|\psi^{(1)}(t)\rangle + \lambda^2|\psi^{(2)}(t)\rangle + \cdots \tag{7.29}$$

and require

$$|\psi^{(1)}(t_0)\rangle = |\psi^{(2)}(t_0)\rangle = \cdots = 0 \tag{7.30}$$

We assume that the interaction is turned on at a time $t > t_0$ so $|\psi^{(0)}(t_0)\rangle = |\psi(t_0)\rangle$ and all other $|\psi^{(i)}(t_0)\rangle = 0$ for $i = 1, 2, 3, \cdots$. Substituting Equation (7.29) in Equation (7.26) we get

$$\begin{aligned} &-\frac{1}{i}\frac{\partial|\psi^{(0)}\rangle}{\partial t} - \frac{1}{i}\lambda\frac{\partial|\psi^{(1)}\rangle}{\partial t} - \frac{1}{i}\lambda^2\frac{\partial|\psi^{(2)}\rangle}{\partial t} - \cdots \\ &= \lambda H_1|\psi^{(0)}(t)\rangle + \lambda^2 H_1|\psi^{(1)}(t)\rangle + \cdots \end{aligned} \tag{7.31}$$

Equating terms of the same order of λ on the left and right hand side of Equation (7.31) we get

$$-\frac{1}{i}\frac{\partial|\psi^{(0)}(t)\rangle}{\partial t} = 0 \tag{7.32}$$

$$-\frac{1}{i}\frac{\partial|\psi^{(1)}(t)\rangle}{\partial t} = H_1(t)\,|\psi^{(0)}(t)\rangle \tag{7.33}$$

$$-\frac{1}{i}\frac{\partial|\psi^{(2)}(t)\rangle}{\partial t} = H_1(t)\,|\psi^{(1)}(t)\rangle \tag{7.34}$$

$$\cdots \tag{7.35}$$

and so on for larger powers of λ. A clear structure can be seen in Equation (7.32) through Equation (7.35). As in Section 1.3 we must employ a bootstrap method to solve this set of coupled equations because an equation of order λ^{n+1} requires the preceding equation of order λ^n to be solved first. We will not go beyond Equation (7.34), second order in λ.

Solving Equation (7.32) we find that $|\psi^{(0)}\rangle$ is constant and

$$|\psi^{(0)}\rangle = |\psi(t_0)\rangle \tag{7.36}$$

using the initial condition in Equation (7.28) and Equation (7.30). Integrating Equation (7.33) we obtain

$$|\psi^{(1)}(t)\rangle = -i \int_{t_0}^{t} dt_1 H_1(t_1)\, |\psi^{(0)}\rangle \tag{7.37}$$

$$= -i \int_{t_0}^{t} dt_1 H_1(t_1)\, |\psi(t_0)\rangle \tag{7.38}$$

where we used Equation (7.36) for $|\psi^{(0)}\rangle$. Integrating Equation (7.34) we obtain

$$|\psi^{(2)}(t)\rangle = -i \int_{t_0}^{t} dt_1 H_1(t_1)\, |\psi^{(1)}(t)\rangle \tag{7.39}$$

$$= (-i)^2 \int_{t_0}^{t} dt_1 H_1(t_1) \int_{t_0}^{t_1} dt_2 H_1(t_2)\, |\psi(t_0)\rangle \tag{7.40}$$

where we used Equation (7.38) for $|\psi^{(1)}(t)\rangle$ in Equation (7.39).

Substituting Equation (7.36), Equation (7.38) and Equation (7.40) in Equation (7.29) and factoring out the time-independent $|\psi(t_0)\rangle$ we obtain

$$|\psi(t)\rangle = \left[1 + (-i) \int_{t_0}^{t} dt_1 H_1(t_1) + (-i)^2 \int_{t_0}^{t} dt_1 H_1(t_1) \int_{t_0}^{t_1} dt_2 H_1(t_2) + \cdots \right] |\psi(t_0)\rangle \tag{7.41}$$

where the parameter λ has been canceled. This is the solution we seek. Its structure is clear and its extension to higher order terms is clear too.

As in Section 1.3 we consider the initial state $|i\rangle$ to be prepared long before the interaction takes place, so we let $t_0 \to -\infty$ and set $|\psi(t_0)\rangle = |i\rangle$. Likewise the final state will be observed long after the interaction has taken place, so we let $t \to +\infty$ and project out the final state of interest $|f\rangle$ from $|\psi(t)\rangle$ by forming the quantity $\langle f|\psi(t)\rangle = S_{fi}$, where S_{fi} is the transition amplitude or S matrix element. Thus we find

$$S_{fi} = \langle f| \left[1 + (-i) \int_{-\infty}^{+\infty} dt_1 H_1(t_1) + (-i)^2 \int_{-\infty}^{+\infty} dt_1 H_1(t_1) \int_{-\infty}^{t_1} dt_2 H_1(t_2) + \cdots \right] |i\rangle \tag{7.42}$$

Consider the third term in the square brackets in Equation (7.42). We would like the upper limit of the integration of the second integral to be $+\infty$ instead of t_1. Changing the limit in that manner doubles the integral as can be shown as follows. The integration limits in Equation (7.42) require that $t_2 \leq t_1$ so in the t_1, t_2 plane we integrate over a triangle limited by the line $t_2 = t_1$ with $t_2 \leq t_1$. The integral is unchanged if we rename the integration variables as they are dummy variables. We replace the integration variable t_1 by t_2 and the integration variable t_2 by t_1. The third term in Equation (7.42) now reads

$$(-i)^2 \int_{-\infty}^{+\infty} dt_2 H_1(t_2) \int_{-\infty}^{t_2} dt_1 H_1(t_1) \Big] \tag{7.43}$$

where the integration variables now satisfy the inequality $t_1 \leq t_2$ so in the t_1, t_2 plane we integrate over a triangle limited by the line $t_2 = t_1$ with $t_2 \geq t_1$. The two triangular integration regions combined form a square, but the integrands in the third term in Equation (7.42) and Equation (7.43) are not the same. We note though that in each case the H_1 at later time is to the left of the H_1 at the earlier time. We can make the integrands the same by writing them as $T[H_1(t_1)H_1(t_2)]$ where the time-ordered product is defined as

$$T[H_1(t_1)H_1(t_2)] = \begin{cases} H_1(t_1)H_1(t_2) & \text{for } t_1 > t_2 \\ H_1(t_2)H_1(t_1) & \text{for } t_1 < t_2 \end{cases} \tag{7.44}$$

The operator evaluated at the latest time is on the left. This definition holds independent of whether $H_1(t_1)$ and $H_1(t_2)$ commute.

Using the Hamiltonian density $\mathcal{H}$ defined in Equation (7.1) we can write the S matrix as

$$\begin{aligned} S = 1 + (-i) \int_{-\infty}^{+\infty} d^4x_1\, \mathcal{H}_I(x_1) \\ + \frac{(-i)^2}{2} \int_{-\infty}^{+\infty} d^4x_1 \int_{-\infty}^{+\infty} d^4x_2\, T[\mathcal{H}_I(x_1)\mathcal{H}_I(x_2)] \end{aligned} \tag{7.45}$$

where we have truncated the perturbative expansion after the second-order term.

As a first application we discuss W and Z decays as these can be treated using first-order perturbation theory. Almost all aspects of doing the calculation of a decay or a scattering process will be encountered. The methodology can be carried over into calculations that use second-order perturbation theory.

7.2 W AND Z BOSON DECAYS

7.2.1 Amplitude

As an example, we evaluate the decay of a W^+ into an ℓ^+ and a ν_ℓ that are members of the same doublet

$$W^+ \to \ell^+ \nu_\ell \tag{7.46}$$

This is a charged weak current interaction described by Equation (7.6). Without loss of generality we will perform the calculation in the W center-of-mass. Thus, we assume that the W has four-momentum $p = (0, im)$ and polarization vector $\varepsilon_t(\mathbf{p})$. Doing the calculation in the W center-of-mass allows for an important simplification in the sum over the direction of the polarization vector of the W^+, necessary for the calculation. We specify the momentum and spin index of the ℓ^+ by $\mathbf{p}_1$ and r, and of the ν_ℓ by $\mathbf{p}_2$ and s. Their masses are m_1 and m_2 respectively. Their four-momenta are $(\mathbf{p}_1, iE_1)$ and $(\mathbf{p}_2, iE_2)$. We must evaluate the S-matrix between the initial state $|i\rangle = |\mathbf{p}, t\rangle$ and the final state $|f\rangle = |\mathbf{p}_1, r; \mathbf{p}_2, s\rangle$ and we must annihilate a W^+ and create an ℓ^+ and a ν_ℓ with specified momentum and spin. The first term in Equation (7.45) obviously does not contribute because $\langle f|1|i\rangle = 0$. The second term in Equation (7.45) will contribute because when using the first term of Equation (7.6) we obtain

$$\langle \ell^+ \nu_\ell | S | W^+ \rangle = -i \int_{-\infty}^{+\infty} d^4x \, \langle \mathbf{p}_1, r; \mathbf{p}_2, s | \, \overline{\psi}(g_V \gamma_\mu + g_A \gamma_\mu \gamma_5) \psi \; W_\mu^\dagger \, |\mathbf{p}, t\rangle \tag{7.47}$$

The creation and annihilation operators work on the ket $|\mathbf{p}, t\rangle$ which can also be written as $|\mathbf{p}, t; 0\rangle$ where the 0 indicates the absence of Dirac particles in the initial state. Likewise, the bra $\langle \mathbf{p}_1, r; \mathbf{p}_2, s|$ can be written as $\langle \mathbf{p}_1, r; \mathbf{p}_2, s; 0|$ where the 0 indicates the absence of a W in the final state. The creation and annihilation operators in the expressions for $\overline{\psi}$, ψ and W_μ work only on the bra and ket; they do not act on the γ matrices or on the exponentials. Those operators can be passed through the gamma matrices, exponentials and polarization vectors at will. When we work out the product of $\overline{\psi}$, ψ and W in Equation (7.47) we obtain eight terms. Only one of these contributes to the process of interest here. We know from the discussion below Equation (6.98) that $W^\dagger$ can annihilate a W^+ (or create a W^-). Likewise, from the discussion below Equation (6.3), ψ can create an ℓ^+ (or annihilate an ℓ^-) while $\overline{\psi}$ can create an ν_ℓ (or annihilate an $\overline{\nu}_\ell$). The creation and annihilation operators appear in sums over momenta and spin indices, and the quantum numbers of the particles to be created or annihilated are guaranteed to appear in the sum because the sum is over all possible values of momenta and spin indices of the system. Of the sums, only terms with the annihilation or creation operators that have the required momenta and spin indices contribute. Thus the W^+, represented by $|\mathbf{p}_1, t\rangle$, is annihilated by the term with $b_t(\mathbf{p})$ giving $b_t(\mathbf{p})|\mathbf{p}, t\rangle = |0\rangle$. The ℓ^+, represented by $\langle \mathbf{p}_1, r|$, is created with the term with $b_r^\dagger(\mathbf{p}_1)$, and the ν_ℓ, represented by $\langle \mathbf{p}_2, s|$, is created with the term with $a_s^\dagger(\mathbf{p}_2)$, giving $a_s^\dagger(\mathbf{p}_2) b_r^\dagger(\mathbf{p}_1)|0\rangle = |\mathbf{p}_1, r; \mathbf{p}_2, s\rangle$. After evaluating the products of the bras and kets (they give 1) we find

$$W_\mu^\dagger |\mathbf{p}, t\rangle = \frac{1}{\sqrt{V}} \frac{1}{\sqrt{2m}} e^{ipx} [\varepsilon_t(\mathbf{p})]_\mu |0\rangle \tag{7.48}$$

and

$$\langle \mathbf{p}_1, r; \mathbf{p}_2, s | \overline{\psi}\, \mathcal{O}_\mu \psi | \mathbf{p}, t\rangle$$
$$= \langle 0 | \frac{1}{\sqrt{V}} \sqrt{\frac{m_2}{E_2}} e^{-ip_2 x}\, \overline{u}_s(\mathbf{p}_2)\, \mathcal{O}_\mu \frac{1}{\sqrt{V}} \sqrt{\frac{m_1}{E_1}} e^{-ip_1 x}\, v_r(\mathbf{p}_1) \tag{7.49}$$

where we have set $g_V \gamma_\mu + g_A \gamma_\mu \gamma_\nu = \mathcal{O}_\mu$. The inner product of the bra $\langle \mathbf{p}_1, r; \mathbf{p}_2, s; 0|$ and the ket $|\mathbf{p}_1, r; \mathbf{p}_2, s; 0\rangle$ that results from the creation and annihilation operators working on $|\mathbf{p}, t; 0\rangle$ equals 1.

The order of spinors and gamma matrices must be respected. Using Equation (7.48) and Equation (7.49) in Equation (7.47) we get

$$\langle \ell^+ \nu_\ell | S | W^+ \rangle = -i \left(\frac{1}{\sqrt{V}} \right)^3 \sqrt{\frac{m_1 m_2}{2m\, E_1 E_2}}\, \overline{u}_s(\mathbf{p}_2)\, \mathcal{O}_\mu v_r(\mathbf{p}_1)\, [\varepsilon_t(\mathbf{p})]_\mu$$
$$\times \int_{-\infty}^{\infty} d^4x\, e^{i(p - p_1 - p_2)x} \tag{7.50}$$

where we used that $\langle 0|0\rangle = 1$. This expression is seen to be a complex number and is the amplitude for the process Equation (7.46) with the W^+ in spin state t and the ℓ^+ and ν_ℓ in spin states r and s respectively. Of course the index μ is understood to be summed over.

In the next section we will introduce a graphical method due to Feynman that helps greatly in writing down amplitudes like Equation (7.50) for an arbitrary process.

The integral over $d^4\mathbf{x}$ can be related to a four-dimensional Dirac delta function

$$\int_{-\infty}^{+\infty} e^{i(p - p_1 - p_2)x} = (2\pi)^4\, \delta^4(p - p_1 - p_2) \tag{7.51}$$

as in Equation (1.98). This relation enforces energy-momentum conservation.

7.2.2 Decay Rate

A decay rate (transition probability per unit time) is obtained using Equation (1.104) with w_{fi} given by

$$w_{fi} = \int dw_{fi} \tag{7.52}$$

and dw_{fi} by

$$dw_{fi} = \frac{1}{T}|\langle f|S|i\rangle|^2 \frac{Vd^3\mathbf{p}_1}{(2\pi)^3}\frac{Vd^3\mathbf{p}_2}{(2\pi)^3} \tag{7.53}$$

The factor $1/T$ accounts for the fact that we want the transition probability per unit time, and T is the time interval in which the decay takes place. T will cancel in the calculation and that is desirable. The second factor is the amplitude $\langle f|S|i\rangle$ squared to get the probability. The last two factors are phase space factors, see Equation (1.119), one for each final state particle. To get the decay rate we integrate over $\mathbf{p}_1$ and $\mathbf{p}_2$ and sum over the spins of the final state particles. We average over the spin of the W^+ in the initial state. In the present case this means summing over r, s and t and a pre-factor 1/3 for the average over the W^+ spin direction.

The square of the amplitude is the product of the amplitude (with a summation over μ) and its complex conjugate (where we change the summation index from μ to ν). The square of the four-dimensional Dirac delta function is real and is evaluated analogously to the square of a one-dimensional Dirac delta function, see Equation (1.102)

$$[\delta^4(p)]^2 = \delta^4(p)\frac{VT}{(2\pi)^4} \tag{7.54}$$

where V is the volume of the box in which the physical system is quantized. We should take the limits $V, T \to \infty$ but V and T cancel, compare the cancelation of T in Equation (1.103). We see that the factor T in Equation (7.54) cancels the factor T in Equation (7.53). The factors V will cancel because we have three in the numerator, one from Equation (7.54) and two from Equation (7.53), and these cancel the three factors of V in the denominator of $|\langle f|S|i\rangle|^2$ of Equation (7.50).

The complex conjugate of a complex number that itself is the product of spinors, γ matrices, and polarization vectors that appear in Equation (7.50) equals the complex conjugate of that complex number. But the Hermitian conjugate of a product of spinors and matrices equals the product of the Hermitian conjugates of the spinors and matrices in reverse order. Thus the complex conjugate of $\bar{u}_s(\mathbf{p}_2)\mathcal{O}_\mu v_r(\mathbf{p}_1)$ is $v_r^\dagger(\mathbf{p}_1)\mathcal{O}_\mu^\dagger\gamma_4 u_s(\mathbf{p}_2)$ which equals $\bar{v}_r(\mathbf{p}_1)\gamma_4\mathcal{O}_\mu^\dagger\gamma_4 u_s(\mathbf{p}_2)$. In the present case $\mathcal{O}_\mu^\dagger = [g_V\gamma_\mu + g_A\gamma_\mu\gamma_5]^\dagger = [g_V^*\gamma_\mu - g_A^*\gamma_\mu\gamma_5]$ where we have used that the five gamma matrices are

Hermitian and we commuted γ_5 with its neighboring γ_μ on the left, using that $\{\gamma_\mu, \gamma_5\} = 0$, see Equation (7.10).

The square of the amplitude also involves $[\varepsilon_t(\mathbf{p})]_\mu\, [\varepsilon_t^*(\mathbf{p})]_\nu$, summed over t, and this sum is given by Equation (6.87). We showed just below Equation (6.88) that this sum is zero when either μ or ν (or both) equal 4. We use this fact in evaluating $\gamma_4 \mathcal{O}_\mu^\dagger \gamma_4$ when we commute the second γ_4 matrix with the γ_5 and the two γ_μ on its left until it is adjacent to the first γ_4, causing both γ_4 to vanish because $\gamma_\mu^2 = 1$. When doing this we use that $\{\gamma_\mu, \gamma_5\} = 0$ and that γ_4 anti-commutes with γ_μ because the contribution with $\mu = 4$ vanishes. We therefore obtain $-[g_V^* \gamma_\mu + g_A^* \gamma_\mu \gamma_5]$ and the $+$ sign between the terms in the brackets is restored, but an overall $-$ sign appears in front. Altogether we get

$$\sum_{r,s,t} [\overline{u}_s(\mathbf{p}_2)\, \mathcal{O}_\mu v_r(\mathbf{p}_1)]\, [\varepsilon_t(\mathbf{p})]_\mu\, [\overline{u}_s(\mathbf{p}_2)\, \mathcal{O}_\nu v_r(\mathbf{p}_1)]^*\, [\varepsilon_t^*(\mathbf{p})]_\nu$$

$$= -\sum_{r,s} [\overline{u}_s(\mathbf{p}_2)\, (g_V \gamma_\mu + g_A \gamma_\mu \gamma_5) v_r(\mathbf{p}_1)] \left[\delta_{\mu\nu} + \frac{p_\mu p_\nu}{m^2}\right]$$

$$\cdot\, [\overline{v}_r(\mathbf{p}_1)\, (g_V^* \gamma_\nu + g_A^* \gamma_\nu \gamma_5) u_s(\mathbf{p}_2)] \qquad (7.55)$$

We discuss next the summation over r and s.

7.2.3 Summation over Spin

The expression $v_r(\mathbf{p}_1)\overline{v}_r(\mathbf{p}_1)$ in Equation (7.55) is a 4×4 matrix because it is the product of a column vector and a row vector. Using the expressions Equation (4.101) and Equation (4.107) for the spinors u and v with $N^2 = (E + m)/(2m)$, one can show that

$$\sum_{i=1}^{2} u_i(\mathbf{p})\overline{u}_i(\mathbf{p}) = \frac{-i\not{p} + m}{2m} \qquad \sum_{i=1}^{2} v_i(\mathbf{p})\overline{v}_i(\mathbf{p}) = \frac{-i\not{p} - m}{2m} \qquad (7.56)$$

with $\not{p} = \gamma_\mu p_\mu$. It is customary to define the operators $\Lambda^\pm(\mathbf{p})$ as

$$\Lambda^\pm(\mathbf{p}) = \frac{\mp i\not{p} + m}{2m} \qquad (7.57)$$

These matrices have uses beyond the present calculation. With Equation (7.57) we can write Equation (7.56) as

$$\sum_{i=1}^{2} u_i(\mathbf{p})\overline{u}_i(\mathbf{p}) = \Lambda^+(\mathbf{p}) \qquad \sum_{i=1}^{2} v_i(\mathbf{p})\overline{v}_i(\mathbf{p}) = -\Lambda^-(\mathbf{p}) \qquad (7.58)$$

Using Equation (7.58) in Equation (7.55) we get

$$\mathcal{I}_{\mu\nu} = -\sum_{r,s}[\overline{u}_s(\mathbf{p}_2)\,(g_V\gamma_\mu + g_A\gamma_\mu\gamma_5)v_r(\mathbf{p}_1)]\,[\overline{v}_r(\mathbf{p}_1)\,(g_V^*\gamma_\nu + g_A^*\gamma_\nu\gamma_5)u_s(\mathbf{p}_2)]$$

$$= -\sum_{s}\overline{u}_s(\mathbf{p}_2)\,(g_V\gamma_\mu + g_A\gamma_\mu\gamma_5)(-\Lambda^-(\mathbf{p}_1))\,(g_V^*\gamma_\nu + g_A^*\gamma_\nu\gamma_5)u_s(\mathbf{p}_2) \tag{7.59}$$

where we introduced the abbreviation $\mathcal{I}$. We will cancel the two minus signs. This expression has the form $[\overline{u}_s(\mathbf{p}_2)]_\alpha A_{\alpha\beta}[u_s(\mathbf{p}_2)]_\beta$ where the summation index s appears only in the two places shown and $A_{\alpha\beta}$ is a 4×4 matrix. To use Equation (7.58) for the sum over s we want $\overline{u}_s(\mathbf{p}_2)$ and $u_s(\mathbf{p}_2)$ adjacent to each other. Because the matrix $A_{\alpha\beta}$ does not contain operators that affect $[u_s(\mathbf{p}_2)]_\beta$, we can move $[u_s(\mathbf{p}_2)]_\beta$ to the left to get $[u_s(\mathbf{p}_2)]_\beta[\overline{u}_s(\mathbf{p}_2)]_\alpha A_{\alpha\beta}$. Therefore

$$\sum_{s}[\overline{u}_s(\mathbf{p}_2)]_\alpha A_{\alpha\beta}[u_s(\mathbf{p}_2)]_\beta = [\Lambda^+(\mathbf{p})]_{\beta\alpha}A_{\alpha\beta} = \mathrm{Tr}[\Lambda^+(\mathbf{p})A] \tag{7.60}$$

where we used Equation (7.58). Using this, Equation (7.59) becomes

$$\mathcal{I}_{\mu\nu} = \mathrm{Tr}[\Lambda^+(\mathbf{p}_2)(g_V\gamma_\mu + g_A\gamma_\mu\gamma_5)\Lambda^-(\mathbf{p}_1)\,(g_V^*\gamma_\nu + g_A^*\gamma_\nu\gamma_5)] \tag{7.61}$$

This expression depends upon the indices μ and ν. We can write $g_V\gamma_\mu + g_A\gamma_\mu\gamma_5 = \gamma_\mu(g_V + g_A\gamma_5)$ and $g_V^*\gamma_\nu + g_A^*\gamma_\nu\gamma_5 = \gamma_\nu(g_V^* + g_A^*\gamma_5)$ and commute the $g_V + g_A\gamma_5$ factor with its neighboring $\Lambda^-(\mathbf{p}_1)\gamma_\nu$ to the right so that it is adjacent to $g_V^* + g_A^*\gamma_5$. With this Equation (7.61) becomes

$$\mathcal{I}_{\mu\nu} = \mathrm{Tr}[\Lambda^+(\mathbf{p}_2)\gamma_\mu\Lambda^-(\mathbf{p}_1)\gamma_\nu\,(g_V + g_A\gamma_5)(g_V^* + g_A^*\gamma_5)] \tag{7.62}$$

The $g_A\gamma_5$ term changed sign twice and we have dropped the mass term in the numerator of $\Lambda^-(\mathbf{p}_1)$ because the final state particles are relativistic.

The product $(g_V + g_A\gamma_5)(g_V^* + g_A^*\gamma_5)$ can be written as $a + b\gamma_5$ with

$$a = |g_V|^2 + |g_A|^2 \qquad\qquad b = g_Vg_A^* + g_V^*g_A \tag{7.63}$$

It is seen that a and b are real. Then Equation (7.62) becomes

$$\mathcal{I}_{\mu\nu} = \mathrm{Tr}[\Lambda^+(\mathbf{p}_2)\gamma_\mu\Lambda^-(\mathbf{p}_1)\gamma_\nu\,(a + b\gamma_5)] \tag{7.64}$$

This expression can be evaluated using a number of theorems that involve the Trace of γ matrices. It can be seen that Equation (7.64) involves the Traces of products of four, three, and two γ matrices (not counting γ_5) and the Traces of these with an additional γ_5.

The Trace of two γ matrices can be evaluated using Equation (4.23). We see that $\mathrm{Tr}[\gamma_\mu\gamma_\nu + \gamma_\nu\gamma_\mu] = 2\,\mathrm{Tr}[\gamma_\mu\gamma_\nu]$ where we used that the Trace of a sum of matrices is the sum of the Traces of the matrices and that the matrices in the Trace of a product of matrices may be cyclically permuted. The right-hand side of Equation (4.23) can be evaluated as $\mathrm{Tr}[2\delta_{\mu\nu}] = 8\delta_{\mu\nu}$. Therefore

$$\mathrm{Tr}[\gamma_\mu\gamma_\nu] = 4\delta_{\mu\nu} \tag{7.65}$$

The trace of three γ matrices can be evaluated by taking the Trace of $\gamma_5^2\gamma_\alpha\gamma_\beta\gamma_\kappa$ which equals the trace of $\gamma_\alpha\gamma_\beta\gamma_\kappa$ because $\gamma_5^2 = 1$. If we cyclically permute matrices, the Trace of their product does not change so $\mathrm{Tr}[\gamma_5^2\gamma_\alpha\gamma_\beta\gamma_\kappa] = \mathrm{Tr}[\gamma_5\gamma_\alpha\gamma_\beta\gamma_\kappa\gamma_5]$. We can commute the left-most γ_5 matrix with its neighbors to the right and after three permutations this γ_5 appears on the right next to the other γ_5 and because $\gamma_5^2 = 1$ they both vanish. But γ_5 anticommutes with the other three so we get three minus signs and the Trace of three γ matrices equals its negative so it is zero. It is easy to extend this argument to the Trace of any odd number of γ matrices so

$$\mathrm{Tr}[\gamma_\mu \cdots \gamma_\alpha \cdots] = 0 \qquad \text{(odd number of } \gamma \text{ matrices)} \tag{7.66}$$

The Trace of four γ matrices $\mathrm{Tr}[\gamma_\alpha\gamma_\beta\gamma_\mu\gamma_\nu]$ can be evaluated by anticommuting γ_ν with its neighbors to the left using Equation (4.23). This gives

$$\begin{aligned}
\gamma_\alpha\gamma_\beta\gamma_\mu\gamma_\nu &= -\gamma_\alpha\gamma_\beta\gamma_\nu\gamma_\mu + \gamma_\alpha\gamma_\beta\, 2\delta_{\mu\nu} \\
&= \gamma_\alpha\gamma_\nu\gamma_\beta\gamma_\mu - \gamma_\alpha\gamma_\mu\, 2\delta_{\beta\nu} + \gamma_\alpha\gamma_\beta\, 2\delta_{\mu\nu} \\
&= -\gamma_\nu\gamma_\alpha\gamma_\beta\gamma_\mu + \gamma_\beta\gamma_\mu\, 2\delta_{\alpha\nu} - \gamma_\alpha\gamma_\mu\, 2\delta_{\beta\nu} + \gamma_\alpha\gamma_\beta\, 2\delta_{\mu\nu}
\end{aligned} \tag{7.67}$$

Taking the Trace of Equation (7.67) and using that $\mathrm{Tr}[\gamma_\alpha\gamma_\beta\gamma_\mu\gamma_\nu] = \mathrm{Tr}[\gamma_\nu\gamma_\alpha\gamma_\beta\gamma_\mu]$ and using Equation (7.65) we find

$$\mathrm{Tr}[\gamma_\alpha\gamma_\beta\gamma_\mu\gamma_\nu] = 4(\delta_{\alpha\beta}\delta_{\mu\nu} - \delta_{\alpha\mu}\delta_{\beta\nu} + \delta_{\alpha\nu}\delta_{\beta\mu}) \tag{7.68}$$

We also need the Trace of products of γ matrices with a γ_5 in the product. Starting with $\mathrm{Tr}[\gamma_\mu\gamma_\nu\gamma_5]$, we note that $\mu \neq \nu$ because if $\mu = \nu$ the Trace becomes $\mathrm{Tr}[\gamma_5]$ and $\mathrm{Tr}[\gamma_5] = 0$, see Equation (7.8). To evaluate $\mathrm{Tr}[\gamma_\mu\gamma_\nu\gamma_5] = \mathrm{Tr}[\gamma_\mu\gamma_\nu\gamma_1\gamma_2\gamma_3\gamma_4]$ we select one of the four γ_1 though γ_4 matrices on the right that has the index ν. We anticommute it with its neighbors on its left until it is adjacent to the other γ_ν. The anticommutation gives a number of minus signs and the product of the two γ_ν equals 1 and can be left out. The Trace now has γ_μ and three of the four γ_1 though γ_4 matrices left. Because $\mu \neq \nu$, one of the latter three has the index μ. We anticommute this one with the γ matrices on its left until it is adjacent to the other γ_μ. We get additional minus signs and the product of the two γ_μ

equals 1 and can also be left out. We are now left with the Trace of two γ matrices whose indices are different. This Trace is seen to be zero from Equation (7.65) so we have

$$\mathrm{Tr}[\gamma_\mu \gamma_\nu \gamma_5] = 0 \tag{7.69}$$

We also need $\mathrm{Tr}[\gamma_\alpha \gamma_\beta \gamma_\mu \gamma_\nu \gamma_5]$. All indices α, β, μ, ν must be different from each other because if two of them are the same we can anticommute their corresponding γ matrices with their neighbors until they are adjacent to each other. The anticommutation gives a number of minus signs and their product being 1, they can be left out. The resulting Trace is of the form of Equation (7.69) and is therefore zero. With all four indices in the Trace different from each other we have for example $\mathrm{Tr}[\gamma_1\gamma_2\gamma_3\gamma_4\gamma_5]$ and this equals $\mathrm{Tr}[\gamma_5\gamma_5]$ which equals $\mathrm{Tr}[1] = 4$. If the four different indices are not in sequential order we can commute their corresponding γ matrices until they are in sequential order. This results in a number of minus signs equal to the number of permutations required. The Levy-Civita symbol $\varepsilon_{\alpha\beta\mu\nu}$ can do this bookkeeping task for us. It is defined to be antisymmetric under exchange of any two adjacent indices and $\varepsilon_{1234} = 1$. We summarize the above results by setting

$$\mathrm{Tr}[\gamma_\alpha \gamma_\beta \gamma_\mu \gamma_\nu \gamma_5] = 4\varepsilon_{\alpha\beta\mu\nu} \tag{7.70}$$

Armed with the properties of Traces given by Equation (7.65) through Equation (7.70) we now return to the evaluation of the Trace in Equation (7.64)

$$\begin{aligned}\mathcal{I}_{\mu\nu} &= \mathrm{Tr}[\Lambda^+(\mathrm{p}_2)\gamma_\mu \Lambda^-(\mathrm{p}_1)\gamma_\nu (a + b\gamma_5) \\ &= \mathrm{Tr}\left[\frac{-i\not{p}_2}{2m_2}\gamma_\mu \frac{i\not{p}_1}{2m_1}\gamma_\nu(a + b\gamma_5)\right] && (7.71)\\ &= \frac{a}{4m_1m_2}\mathrm{Tr}[\not{p}_2\gamma_\mu\not{p}_1\gamma_\nu] + \frac{b}{4m_1m_2}\mathrm{Tr}[\not{p}_2\gamma_\mu\not{p}_1\gamma_\nu\gamma_5] && (7.72)\end{aligned}$$

We used Equation (7.57) for $\Lambda^\pm$ in Equation (7.71), dropped the mass terms in the numerators, and multiplied out the factors with the parentheses. The two terms can be evaluated using the Trace theorems Equation (7.65) through Equation (7.70). We evaluate the first term in detail.

$$\begin{aligned}\mathrm{Tr}[\not{p}_2\gamma_\mu\not{p}_1\gamma_\nu] &= \mathrm{Tr}[\gamma_\alpha\gamma_\mu\gamma_\beta\gamma_\nu]\, p_{2\alpha}p_{1\beta} && (7.73)\\ &= 4(\delta_{\alpha\mu}\delta_{\beta\nu} - \delta_{\alpha\beta}\delta_{\mu\nu} + \delta_{\alpha\nu}\delta_{\mu\beta})\, p_{2\alpha}p_{1\beta} && (7.74)\\ &= 4[p_{1\mu}p_{2\nu} + p_{2\mu}p_{1\nu} - (p_1p_2)\,\delta_{\mu\nu}] && (7.75)\end{aligned}$$

We introduced the dummy summation indices α and β in Equation (7.73) and used Equation (7.68) for the Trace of the four γ matrices to obtain Equation (7.74). The result in Equation (7.75) is seen to be symmetric under exchange of the indices μ and ν and symmetric under exchange of the indices 1 and 2. Evaluating the second term in Equation (7.72) in the same manner, we obtain using Equation (7.70)

$$\mathcal{I}_{\mu\nu} = \frac{a}{m_1 m_2}[p_{1\mu}p_{2\nu} + p_{2\mu}p_{1\nu} - (p_1 p_2)\,\delta_{\mu\nu}] + \frac{b}{m_1 m_2}\varepsilon_{\mu\nu\alpha\beta}p_{1\alpha}p_{2\beta} \tag{7.76}$$

We exchanged the dummy summation indices α and β in the second term and then moved the indices μ and ν to the front to get the four indices in the order shown. The second term in this expression is seen to be antisymmetric under exchange of the indices μ and ν and antisymmetric under the simultaneous exchange of the indices α with β and the indices 1 with 2.

The result in Equation (7.76) must now be contracted with $\delta_{\mu\nu} + p_\mu p_\nu/m^2$, see Equation (7.55), an expression that is symmetric under exchange of the indices μ and ν. When we contract an expression that is symmetric in the indices μ and ν with an expression that is antisymmetric in the indices μ and ν, the result is identically zero, see Equation (4.89). So the term proportional to b drops out.

The result of the contraction of Equation (7.76) with $\delta_{\mu\nu} + p_\mu p_\nu/m^2$ is

$$\mathcal{I}_{\mu\nu}\left[\delta_{\mu\nu} + \frac{p_\mu p_\nu}{m^2}\right] = \frac{a}{m_1 m_2}\left[(p_1 p_2) - \frac{2(p_1 p)(p_2 p)}{m^2}\right] \tag{7.77}$$

where we used $\delta_{\mu\nu}\delta_{\mu\nu} = 4$.

Using Equation (7.77) in Equation (7.55) we get

$$\sum_{r,s,t}[\overline{u}_s(\mathbf{p}_2)\,\mathcal{O}_\mu v_r(\mathbf{p}_1)]\,[\overline{u}_s(\mathbf{p}_2)\,\mathcal{O}_\mu v_r(\mathbf{p}_1)]^*[\varepsilon_t(\mathbf{p})]_\mu\,[\varepsilon_t^*(\mathbf{p})]_\nu$$

$$= \frac{a}{m_1 m_2}\left[\frac{2(p_1 p)(p_2 p)}{m^2} - (p_1 p_2)\right] \tag{7.78}$$

Using Equation (7.53) for the transition probability dw_{fi} and Equation (7.50) and Equation (7.51) and the results from the summation over spin indices, we get

$$dw_{fi} = \frac{1}{(2\pi)^2}\frac{1}{2m\,E_1 E_2}\frac{1}{3}a\left[\frac{2(p_1 p)(p_2 p)}{m^2} - (p_1 p_2)\right]$$

$$\times\,\delta^4(p - p_1 - p_2)\,d^3\mathbf{p}_1 d^3\mathbf{p}_2 \tag{7.79}$$

The masses m_1 and m_2 have canceled between the normalization factors of the spinors in the numerator of Equation (7.50) and in the denominator of Equation (7.78). This is good because the neutrino mass is essentially zero so its cancelation is desirable. The cancelation of the V and T was already achieved, see the discussion below Equation (7.54). The factor $(2\pi)^2$ in the denominator results from the square of $(2\pi)^4$ in the numerator which comes from the square of the integral over x in Equation (7.50), see Equation (7.51), from the two factors $(2\pi)^3$ in the denominators of Equation (7.53) and the $(2\pi)^4$ in the denominator in Equation (7.54). The factor 1/3 is from the averaging over the three spin directions of the W. Energy-momentum is conserved owing to the four-dimensional Dirac delta function. Note the symmetry in the indices 1 and 2.

To obtain the total transition probability per unit time we must integrate over the allowed phase space of the two particles in the final state.

7.2.4 Integration over Phase Space

To integrate Equation (7.79) over $\mathbf{p}_1$ and $\mathbf{p}_2$ we first evaluate the expression within the square bracket. Using $p_1p_2 = \mathbf{p}_1 \cdot \mathbf{p}_2 - E_1E_2$, $p_1p = -mE_1$, and $p_2p = -mE_2$ we get that

$$\left[\frac{2(p_1p)(p_2p)}{m^2} - (p_1p_2)\right] = 3E_1E_2 - \mathbf{p}_1 \cdot \mathbf{p}_2 \tag{7.80}$$

Thus we can write Equation (7.79) as

$$dw_{fi} = f(\mathbf{p}_1, \mathbf{p}_2)\,\delta^4(p - p_1 - p_2)\,d^3\mathbf{p}_1 d^3\mathbf{p}_2 \tag{7.81}$$

where $f(\mathbf{p}_1, \mathbf{p}_2)$ includes all factors that multiply $\delta^4(p - p_1 - p_2)\,d^3\mathbf{p}_1 d^3\mathbf{p}_2$ in Equation (7.79). E_1 and E_2 are of course functions of $\mathbf{p}_1$ and $\mathbf{p}_2$.

We integrate Equation (7.81) over $\mathbf{p}_2$ using three of the four Dirac delta functions

$$dw_{fi} = \delta(m - E_1 - E_2)\,d^3\mathbf{p}_1 \int f(\mathbf{p}_1, \mathbf{p}_2)\,\delta^3(-\mathbf{p}_1 - \mathbf{p}_2)\,d^3\mathbf{p}_2 \tag{7.82}$$

In order to use the relation $\int f(x)\,\delta(x - a)dx = f(a)$ we must change the sign of the argument in the three delta functions in Equation (7.82). To do this we use that $\delta(ax) = \delta(x)/a$ to get $\delta(-x) = -\delta(x)$. Changing the sign of the argument of the three delta functions in Equation (7.82) gives three minus signs, so we get

$$\begin{aligned} dw_{fi} &= -\delta(m - E_1 - E_2)\,d^3\mathbf{p}_1 \int f(\mathbf{p}_1, \mathbf{p}_2)\,\delta^3(\mathbf{p}_1 + \mathbf{p}_2)\,d^3\mathbf{p}_2 \\ &= -\delta(m - E_1 - E_2)\,d^3\mathbf{p}_1 f(\mathbf{p}_1, -\mathbf{p}_1) \end{aligned} \tag{7.83}$$

This means that everywhere in $f(\mathbf{p}_1, \mathbf{p}_2)$ we must replace $\mathbf{p}_2$ by $-\mathbf{p}_1$. This makes sense because in the W center-of-mass the two final state particles have momenta equal in length and opposite in direction. E_2 now is equal to $\sqrt{(-\mathbf{p}_1)^2 + m_2^2}$ and in general $E_1 \neq E_2$. Because we have made the approximation that $m_1 = m_2$ (equal zero), $E_1 = E_2$ in the present case.

Next we integrate Equation (7.83) over $\mathbf{p}_1$ using the remaining delta function $\delta(m - E_1 - E_2)$

$$
\begin{aligned}
dw_{fi} &= -\int d^3\mathbf{p}_1 f(\mathbf{p}_1, -\mathbf{p}_1)\delta(m - E_1 - E_2) \\
&= -\int \mathbf{p}_1^2 d|\mathbf{p}_1| d\Omega_1 f(\mathbf{p}_1, -\mathbf{p}_1)\delta(m - E_1 - E_2)
\end{aligned} \tag{7.84}
$$

The integral over the solid angle $d\Omega_1$ can be done immediately, and gives 4π, because there is no angular dependence in the integrand. Because the argument of the delta function is itself a function of the integration variable $|\mathbf{p}_1|$, we use the property

$$
\int f(x)\delta[g(x)]dx = \frac{f(x_0)}{|g'(x_0)|} \tag{7.85}
$$

where x_0 is the solution of $g(x) = 0$ and g' is the derivative of $g(x)$. In the present case $g(|\mathbf{p}_1|) = m - \sqrt{\mathbf{p}_1^2 + m_1^2} - \sqrt{(-\mathbf{p}_1)^2 + m_2^2}$ and the solution of $g(|\mathbf{p}_1|) = 0$ is

$$
|\mathbf{p}_1| = \frac{\sqrt{[m^2 - (m_1 + m_2)^2][m^2 - (m_1 - m_2)^2]}}{2m} \tag{7.86}
$$

$$
= \frac{m}{2} \quad \text{for } m_1 = m_2 = 0 \tag{7.87}
$$

Also

$$
E_1 = \frac{m^2 - m_2^2 + m_1^2}{2m} \tag{7.88}
$$

$$
= \frac{m}{2} \quad \text{for } m_1 = m_2 = 0 \tag{7.89}
$$

$$
E_2 = \frac{m^2 - m_1^2 + m_2^2}{2m} \tag{7.90}
$$

$$
= \frac{m}{2} \quad \text{for } m_1 = m_2 = 0 \tag{7.91}
$$

That $E_1 = |\mathbf{p}_1| = E_2 = |\mathbf{p}_2|$ makes sense in the approximation that $m_1 = m_2 = 0$. The derivative of $g(|\mathbf{p}_1|)$ with respect to $|\mathbf{p}_1|$ is

$$g'(|\mathbf{p}_1|) = -\frac{|\mathbf{p}_1|}{E_1} - \frac{|\mathbf{p}_2|}{E_2} = -\frac{|\mathbf{p}_1|(E_1 + E_2)}{E_1 E_2} \tag{7.92}$$

$$= -2 \quad \text{for } m_1 = m_2 = 0 \tag{7.93}$$

Using Equation (7.93) in Equation (7.84) we get

$$w_{fi} = 4\pi|\mathbf{p}_1|\frac{E_1 E_2}{E_1 + E_2} f(\mathbf{p}_1, -\mathbf{p}_1) \tag{7.94}$$

$$= \frac{\pi}{2} m^2 f(\mathbf{p}_1, -\mathbf{p}_1) \quad \text{for } m_1 = m_2 = 0 \tag{7.95}$$

In the limit $m_1 = m_2 = 0$, the expression in Equation (7.80) equals $4E_1E_2$. Substituting this result in the expression for $f(\mathbf{p}_1, \mathbf{p}_2)$ with $\mathbf{p}_2 = -\mathbf{p}_1$ obtained from a comparison of Equation (7.81) with Equation (7.79) and using Equation (7.87), Equation (7.89) and Equation (7.91) we obtain

$$w_{fi} = \frac{ma}{12\pi} \tag{7.96}$$

This expression has the units of inverse time because of the presence of the factor m in the numerator. It is proportional to coupling constants squared through the factor a, see Equation (7.63). The dependence upon a and m could have been guessed at the onset of the calculation by counting vertices and dimensional analysis respectively. All the calculation is good for is to obtain the numerical prefactor. All that work for a factor 12π in the denominator!

7.2.5 Interpretation

In the case of $W \to \ell^+\nu_\ell$, a charged weak interaction, we have that $g_V = g_A = g_W$ and with Equation (7.63) we find that $a = 2g_W^2$. Substitution in Equation (7.96) gives

$$w_{fi} = \frac{g_W^2 m_W}{6\pi} \tag{7.97}$$

For historical reasons it is customary to use Fermi's coupling constant G_F defined by

$$\frac{G_F}{\sqrt{2}} = \frac{g_W^2}{m_W^2} \tag{7.98}$$

and G_F has units of inverse mass squared (or time squared). With this definition of G_F, Equation (7.97) becomes

$$w_{fi} = \frac{G_F m_W^3}{6\sqrt{2}\pi} \tag{7.99}$$

These expressions are in natural units. Reinstating the proper powers of $\hbar$ and c we find for Equation (7.97)

$$w_{fi} = \frac{g_W^2 m_W c^2}{6\pi \hbar} \tag{7.100}$$

and for Equation (7.99)

$$w_{fi} = \frac{G_F m_W^3 c^2}{6\sqrt{2}\pi \hbar} \tag{7.101}$$

Numerically, $G_F = 1.01 \times 10^{-5}/m_N^2$ with $m_N = 0.931$ GeV, the mass of a nucleon (the average of the proton and neutron mass). Substituting this value for G_F in Equation (7.101) we get

$$w_{fi} = \frac{1.01 \times 10^{-5}\, m_W c^2}{6\sqrt{2}\pi \hbar} \frac{m_W^2}{m_N^2} \tag{7.102}$$

Substituting $m_W = 80.4\,\text{GeV}/c^2, \hbar = 6.58 \times 10^{-25}$ GeVs, we find $w_{fi} = 3.45 \times 10^{23}\,\text{s}^{-1}$. Instead of w_{fi} one finds in the literature the partial width for a decay. It is defined as

$$\Gamma = w_{fi} h \tag{7.103}$$

and is expressed in units of energy. In the present case we get $\Gamma(W \to \ell \nu_\ell) = 227$ MeV, in excellent agreement with the experimental value of 227 MeV.

We can obtain the decay rate for other decay modes of the W and of the Z with a simple substitution of coupling constants and masses. For example, for $Z \to e^+ e^-$ we use Equation (7.96), using for a the result obtained from Equation (7.63) with $g_V = g_Z(\sin^2\theta_W - \frac{1}{4})/\cos\theta_W)$ and $g_A = \frac{1}{4} g_Z/\cos\theta_W$ where θ_W is the weak mixing angle (also called the Weinberg angle). The coupling constant g_Z is related to the the Fermi constant by the relation

$$\frac{G_F}{\sqrt{2}} = \frac{g_Z^2}{8m_W^2} \tag{7.104}$$

Compare with Equation (7.98). We obtain $\Gamma(Z \to e^+ e^-) = 84.0$ MeV, in excellent agreement with the experimental value of 84.1 MeV.

These expressions for g_V and g_A are from the standard electroweak theory. A discussion of this theory is outside the scope of this book. The weak mixing angle is a parameter that was determined experimentally and found to be given by $\sin^2\theta_W \approx 0.23$.

7.3 FEYNMAN GRAPHS

In the previous section we discussed the decay $W^+ \to \ell^+\nu_\ell$. If we are interested instead in $W^- \to \ell^-\overline{\nu}_\ell$ we can find the amplitude in the same way we found the amplitude for the charge conjugated W^+ decay. This time we must use the second term in Equation (7.6) because we must annihilate a W^-. Again we work in the W center-of-mass. The momentum and spin index of the ℓ^- are $\mathbf{p}_1$ and r and those of the $\overline{\nu}_\ell$ are $\mathbf{p}_2$ and s. Their masses are m_1 and m_2 respectively. Their four-momenta are $(\mathbf{p}_1, E_1)$ and $(\mathbf{p}_2, E_2)$. We find the amplitude to be

$$\langle \ell^-\overline{\nu}_\ell|S|W^-\rangle = -i\sqrt{\frac{m_1m_2}{2m\,E_1E_2}}\overline{u}_r(\mathbf{p}_1)\,(g_V\gamma_\mu + g_A\gamma_\mu\gamma_5)v_s(\mathbf{p}_2)[\varepsilon_t(\mathbf{p})]_\mu \times (2\pi)^4\delta^4(p - p_1 - p_2) \tag{7.105}$$

where we have substituted the four-dimensional delta function for the four-dimensional integral over space-time and left out the factors involving the volume V as the latter will cancel. When we compare this amplitude with the amplitude in Equation (7.50) for W^+ decay, we find that the two amplitudes turn into each other under the simultaneous exchange of $\mathbf{p}_1 \leftrightarrow \mathbf{p}_2$ and $r \leftrightarrow s$. The calculation for the decay rate of the $W^- \to \ell^-\overline{\nu}_\ell$ is identical to the calculation for $W^+ \to \ell^+\nu_\ell$ with the indices 1 and 2 exchanged in, for example, Equation (7.79). But this expression is symmetric in the indices 1 and 2, so does not change. The integration over phase space is symmetric under exchange of the indices 1 and 2 and also gives the same result. Therefore the decay rate is the same, see Equation (7.96) and following.

We will consider $e^+\ e^-$ annihilation and $e^-\mu^+$ elastic scattering next. These are electromagnetic processes so we must use the corresponding Hamiltonian density Equation (7.2) with the electromagnetic current given by Equation (7.3) or Equation (7.4) and Equation (7.5). We must annihilate the two particles in the initial state and create the two particles in the final state. Inspection of Equation (7.45) shows that the first two terms are unable to do this and that we must consider the third term with the time-ordered product. While we had eight possible terms when working out Equation (7.47), we will now have 64 terms. It turns out that only two of these contribute. Feynman invented an intuitive graphical method (Feynman graphs) that aids in finding the terms that contribute to a given process. We show in Figure 7.1 the graphical elements that he introduced.

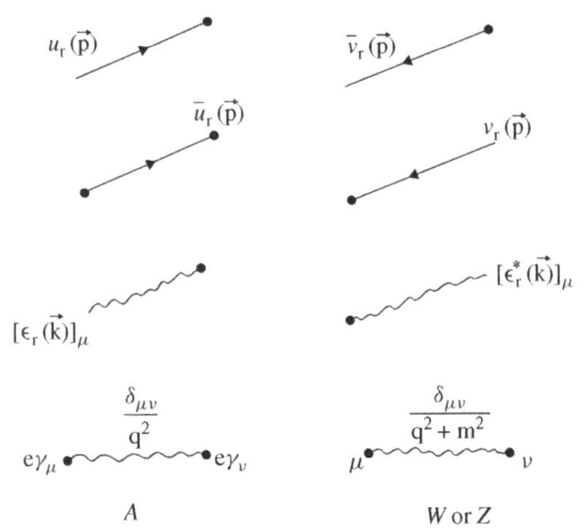

Figure 7.1 Building blocks for the graphical representation of terms in the perturbation series

Each represents a building block representing u, $\bar{v}$, $\bar{u}$, v, A, W or Z. These are used in the graphical representation of terms in the series expansion in Equation (7.45). In general, fermions are represented by solid lines with a dot at one end, and intermediate bosons are represented by wavy lines with a dot at one end. Time is from left to right. The dot represents a space-time point x where a particle is created or annihilated. Arrows distinguish between fermions and anti-fermions, with the convention that anti-fermions run backward in time, so the arrow is to the left for those. Note that the arrows automatically reflect conservation of fermion number. Because there is no conservation of boson number there are no arrows on the wavy lines. Next to each line we write the factor in the series expansion that the line represents. One usually leaves out the normalization factors. The exponential factors lead to four-momentum conservation at each vertex so they may be left out as well. To get all factors 2 and π and the normalization factors right, it is recommended to use the algebraic expression for the amplitude and not try to read them off the graphs using detailed rules that can be formulated. The wavy lines with a dot at each end represent time-ordered products involving intermediate bosons such as the γ, W and Z.

Interactions between fermions and bosons are represented by a triplet of two solid and one wavy line as shown in Figure 7.2: one triplet for each appearance of $\mathcal{H}_I$ in Equation (7.45). So the second term in Equation (7.45) is represented by a single triplet, the third term in Equation (7.45) by two triplets connected by a wavy line, and so on for higher order terms in Equation (7.45). At the vertex of each triplet we write the operator (with

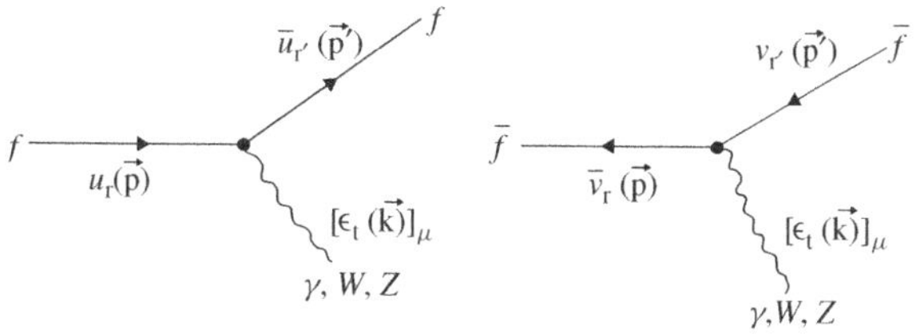

Figure 7.2 Graphical representation of interactions

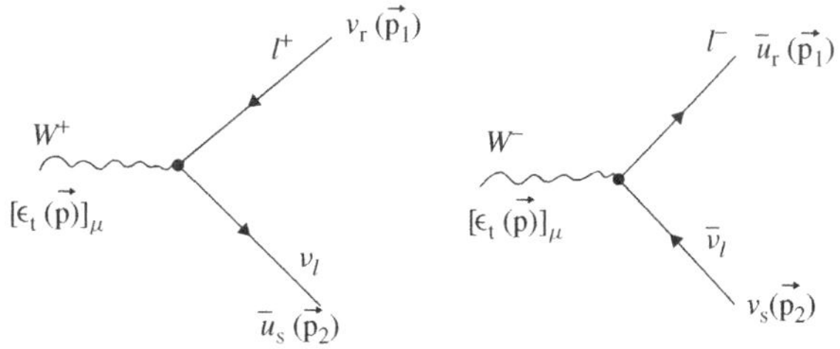

Figure 7.3 Feynman graphs for *W* decays

the associated coupling constant) that is sandwiched between two spinors. Examples are $q_i\gamma_\mu$ for an electromagnetic interaction, and $(g_V\gamma_\mu + g_A\gamma_\mu\gamma_5)$ for a weak interaction. It is easy to see that the two *W* decays we discussed thus far are represented by the two graphs in Figure 7.3.

When writing down the amplitude that corresponds to a graph, one must keep in mind that the amplitude is a scalar, so it must be of the form of a 1×4 row vector times a 4×4 matrix times a 4×1 column vector. This determines the order of the factors seen in graphs.

7.4 SECOND ORDER PROCESSES AND PROPAGATORS

7.4.1 Annihilation and Scattering

As an example we consider $e^+ e^-$ annihilation through an intermediate photon

$$e^+ e^- \to \mu^+\mu^- \tag{7.106}$$

The four-momenta and spin indices are p_1, r and p_2, s for the e^+ and e^- respectively and p'_1, r' and p'_2, s' for the μ^+ and μ^- respectively, and their masses are m_1 and m_2 for the e^+, e^- and μ^+, μ^- respectively. We already noted that this is a second-order process involving the third term in

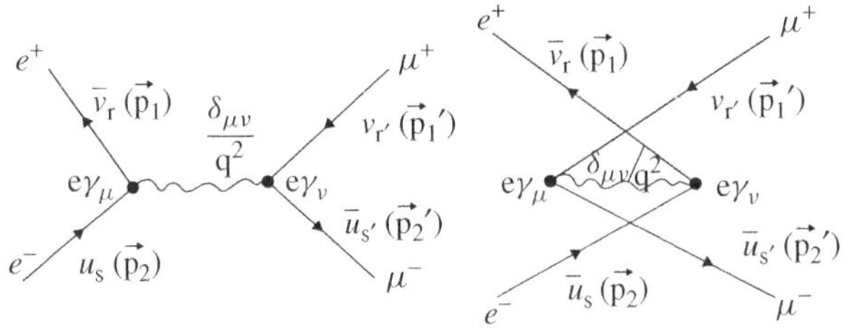

Figure 7.4 Feynman graphs for electron positron annihilation into muon pairs

Equation (7.45). We present its two graphical representations in Figure 7.4. Their amplitudes can be written down with the rules given above and are

$$\bar{v}_r(\mathbf{p}_1)(e\gamma_\mu)\,u_s(\mathbf{p}_2)\langle 0|T[A_\mu(x_1)A_\nu(x_2)]|0\rangle\,\bar{u}_{s'}(\mathbf{p}_2')(e\gamma_\nu)v_{r'}(\mathbf{p}_1') \quad (7.107)$$

and

$$\bar{u}_{s'}(\mathbf{p}_2')(e\gamma_\mu)\,v_{r'}(\mathbf{p}_1')\langle 0|T[A_\mu(x_1)A_\nu(x_2)]|0\rangle\,\bar{v}_r(\mathbf{p}_1)(e\gamma_\nu)\,u_s(\mathbf{p}_2) \quad (7.108)$$

Inspection of the two expressions shows that they are equal to each other if x_1 and x_2 are exchanged, in accordance with the graphs in Figure 7.4. We will show that the time-ordered product $\langle 0|T[A_\mu(x_1)A_\nu(x_2)]|0\rangle$ is symmetric under exchange of x_1 and x_2 and under exchange of μ and ν. Therefore the two amplitudes Equation (7.107) and Equation (7.108) are equal to each other giving a factor two. This factor two cancels the factor two in the denominator of the third term in Equation (7.45). This property is common to all second-order graphs, so it is sufficient to consider only one graph and drop the factor two in Equation (7.45). Note the curious fact that in the second graph the $\mu^+\ \mu^-$ appear to be produced 'before' the $e^+\ e^-$ annihilation. This is quantum physics at work!

Next consider elastic electron-muon scattering

$$e^-\mu^+ \to e^-\mu^+ \quad (7.109)$$

The four-momenta and spin indices are p_1, r and p_2, s for the e^- and μ^+ in the initial state respectively and p_1', r' and p_2', s' for the e^- and μ^+ in the final state respectively, and their masses are m_1 and m_2 for the e^+, e^- and μ^+, μ^- respectively. Also this is a second-order process involving the third term in Equation (7.45). We present its graphical representation in Figure 7.5 where we have omitted a second graph in which x_1 and x_2 are exchanged, because this graph's amplitude is equal to the amplitude of the first graph. The amplitude can be written down with the rules given above and are

$$\bar{u}_{r'}(\mathbf{p}_1')(-e\gamma_\mu)\,u_r(\mathbf{p}_1)\langle 0|T[A_\mu(x_1)A_\nu(x_2)]|0\rangle\,\bar{v}_s(\mathbf{p}_2)(e\gamma_\nu)\,v_{s'}(\mathbf{p}_2') \quad (7.110)$$

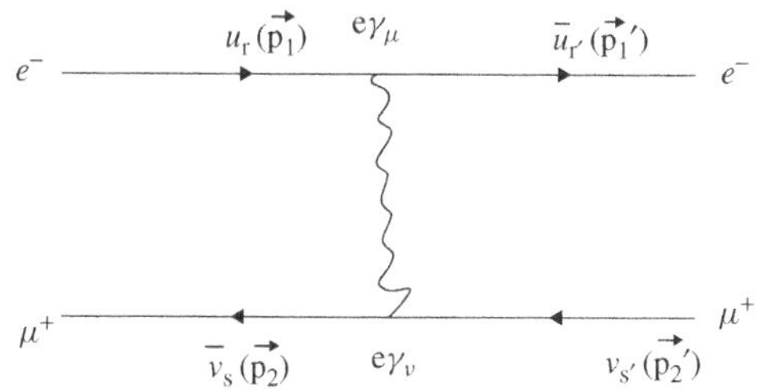

Figure 7.5 Feynman graph for electron-muon scattering

Comparing Equation (7.110) with Equation (7.107) and Equation (7.108), we see that Equation (7.110) can be transformed into Equation (7.107) or Equation (7.108) with the simultaneous changes in Equation (7.110) of $\mathbf{p}_1' \leftrightarrow \mathbf{p}_2$, $\mathbf{p}_1 \leftrightarrow \mathbf{p}_2'$, $r' \leftrightarrow s$, $r \leftrightarrow s'$, and $u \leftrightarrow v$. This will save time when we calculate the square of the amplitude and average of the spins of the initial state particles and sum over the spins of the final state particles.

For $e^-\mu^- \rightarrow e^-\mu^-$ elastic scattering we get the amplitude

$$\bar{u}_{r'}(\mathbf{p}_1')(-e\gamma_\mu)\,u_r(\mathbf{p}_1)\langle 0|T[A_\mu(x_1)A_\nu(x_2)]|0\rangle\,\bar{u}_{s'}(\mathbf{p}_2')(-e\gamma_\nu)\,u_s(\mathbf{p}_2) \quad (7.111)$$

and similar amplitudes can be written down for the other two charge combinations. The amplitudes can be related to each with substitutions of the type above, and again we will save time when we calculate the square of the amplitude and average of the spins of the initial state particles and sum over the spins of the final state particles.

We see from the examples that the time-ordered product $\langle 0|T[A_\mu(x_1)A_\nu(x_2)]|0\rangle$ enters in each of them. The time-ordered product is also called a propagator.

We now turn to its evaluation.

7.4.2 Time-Ordered Product

The time-ordered product $\langle 0|T[A_\mu(x_1)A_\nu(x_2)]|0\rangle$ is defined with the aid of Equation (7.44). It can also be written in the form

$$\begin{aligned}\mathcal{T}_{\mu\nu}(x_1,x_2) &= \langle 0|T[A_\mu(x_1)A_\nu(x_2)]|0\rangle \\ &= \langle 0|A_\mu(x_1)A_\nu(x_2)|0\rangle\theta(t_1-t_2)+\langle 0|A_\nu(x_2)A_\mu(x_1)|0\rangle\theta(t_2-t_1)\end{aligned} \quad (7.112)$$

where θ is the step function that is defined to equal 1 if its argument is positive, and 0 otherwise. Note that the order of the $A_\mu(x_1)$ and $A_\nu(x_2)$

is exchanged without regard to whether they commute or not. When we substitute the plane wave expansion Equation (1.30) in this expression and allow for four components in A instead of three, we find that the only combination of $a_\lambda(\mathbf{k})$ and $a^\dagger_{\lambda'}(\mathbf{k}')$ that gives a non-zero contribution is the combination $a_\lambda(\mathbf{k})a^\dagger_\lambda(\mathbf{k})$ because $a_\lambda(\mathbf{k})|0\rangle = 0$ for all $\mathbf{k}, \lambda$ so we need to create a photon of arbitrary momentum and polarization with $a^\dagger_\lambda(\mathbf{k})$ to obtain $a^\dagger_\lambda(\mathbf{k})|0\rangle = |\mathbf{k}, \lambda\rangle$. This photon must then be annihilated with $a_\lambda(\mathbf{k})$ (note the momentum and polarization) to obtain $a_\lambda(\mathbf{k})|\mathbf{k}, \lambda\rangle = |0\rangle$ so that we are left with the ket $|0\rangle$. The product of the bra $\langle 0|$ and the ket $|0\rangle$ is left and $\langle 0|0\rangle = 1$.

With the factors that belong to $a_\lambda(\mathbf{k})$ and $a^\dagger_\lambda(\mathbf{k})$ we obtain

$$T_{\mu\nu}(x_1, x_2) = \int \frac{d^3\mathbf{k}}{(2\pi)^3} \frac{1}{2\omega_k} \left[e^{ik(x_1 - x_2)} \theta(t_1 - t_2) + e^{ik(x_2 - x_1)} \theta(t_2 - t_1) \right]$$
$$\times \sum_\lambda [\varepsilon^*_\lambda(\mathbf{k})]_\mu [\varepsilon_\lambda(\mathbf{k})]_\nu \tag{7.113}$$

where we have taken common factors and the integral over $\mathbf{k}$ outside the square brackets. The polarization vectors $\varepsilon_\lambda(\mathbf{k})$ are complex numbers and can be put in any order.

We replace the θ-functions by an integral using

$$\theta(t)\, e^{-ibt} = \frac{1}{2\pi i} \int_{-\infty}^{+\infty} dz \frac{e^{-izt}}{b - z - i\varepsilon} \tag{7.114}$$

where t, b, ε and z are real parameters with $b > 0$ and $\varepsilon > 0$. This theorem can be proved using Cauchy's Theorem of integration in the complex plane. Cauchy's Theorem reads

$$f(a) = \frac{1}{2\pi i} \oint \frac{f(z)\, dz}{z - a} \tag{7.115}$$

where a and z are complex numbers. The integral is along a closed path in the complex plane with the positive direction being the left-handed direction (from the positive real axis toward the positive imaginary axis). The integrand in Equation (7.115) has a pole at $z = a$. If the closed path encloses the pole, Equation (7.115) applies. If the closed path does not, the integral on the right-hand side of Equation (7.115) is zero.

We choose a closed path along the real axis from $-R$ to $+R$ and a half circle with radius R in either the lower half of the complex plane if $t > 0$ (and the closed path is followed in the negative direction) or the upper half if $t < 0$ (and the path is followed in the positive direction). With the choice dependent upon the sign of t, the integrand vanishes along the circle because its exponential has a real part multiplying z that is negative.

The integrand in Equation (7.114) has a pole at $z = b - i\varepsilon$ if we allow for the moment z in Equation (7.114) to be complex. This pole has the coordinates $(b, -\varepsilon)$ and thus is located in the complex plane below the positive real axis. If $t > 0$, the pole is included within the closed path and using Equation (7.115) the right-hand side of Equation (7.114) becomes

$$\mathrm{e}^{-i(b-i\varepsilon)t}(-1)^2 \tag{7.116}$$

where the two -1 factors come from the fact that we follow the closed path in the negative direction and that z in the denominator of Equation (7.114) and Equation (7.115) has opposite signs. If $t < 0$, the pole is not enclosed by the path and the right-hand side of Equation (7.114) is zero. Comparison of these results with the left-hand side of Equation (7.114) shows that they are correctly represented if we let $\varepsilon \to 0$.

We use Equation (7.116) with the replacements $b \to \omega_k$ and $z \to \omega$ where $\omega_k = |\mathbf{k}|$ and ω represents the energy associated with the wave-vector $\mathbf{k}$ (but in general $\omega \neq \omega_k$) to obtain

$$\theta(t)\,\mathrm{e}^{-i\omega_k t} = \frac{1}{2\pi i}\int_{-\infty}^{+\infty} d\omega\,\frac{\mathrm{e}^{-i\omega t}}{\omega_k - \omega} \tag{7.117}$$

We note that a photon with energy ω and wave-vector $\mathbf{k}$ is not massless. Using the Einstein relation Equation (4.1), its mass is $\mathbf{k}^2 - \omega^2 \neq 0$. We call such a photon 'of the mass shell'. In general, particles whose momentum and energy do not satisfy the Einstein relation are called 'off the mass shell'.

The parameter t in Equation (7.114) and thus in Equation (7.117) is real and can take on any value between $-\infty$ and $+\infty$. We replace in Equation (7.117) $t \to -t$ to obtain

$$\theta(-t)\,\mathrm{e}^{+i\omega_k t} = \frac{1}{2\pi i}\int_{-\infty}^{+\infty} d\omega\,\frac{\mathrm{e}^{+i\omega t}}{\omega_k - \omega} \tag{7.118}$$

Now set $t = t_1 - t_2$ in Equation (7.117) and Equation (7.118) and substitute the resulting two expressions in Equation (7.113) to obtain

$$T_{\mu\nu}(x_1, x_2) = \int \frac{d^3\mathbf{k}}{(2\pi)^3}\frac{d\omega}{2\omega_k}\left[\frac{\mathrm{e}^{i\mathbf{k}\cdot(\mathbf{x}_1-\mathbf{x}_2)}\mathrm{e}^{-i\omega t}}{\omega_k - \omega} + \frac{\mathrm{e}^{i\mathbf{k}\cdot(\mathbf{x}_2-\mathbf{x}_1)}\mathrm{e}^{+i\omega t}}{\omega_k - \omega}\right]$$
$$\times \sum_\lambda [\varepsilon_\lambda^*(\mathbf{k})]_\mu [\varepsilon_\lambda(\mathbf{k})]_\nu \tag{7.119}$$

We would like to pull out a common exponential $\exp[ik(x_1 - x_2)]$ shown in the first term in the expression in the square brackets. The exponential in the second term in the square brackets can be brought in the same form as the first term with the replacements $\mathbf{k} \to -\mathbf{k}$ and $\omega \to -\omega$. The Jacobian associated with this change of variables is $(-1)^4$. The new integration limits are now from $+\infty$ to $-\infty$. Reversing these gives another $(-1)^4$. We obtain

$$\mathcal{T}_{\mu\nu}(x_1, x_2) = \int \frac{d^4k}{(2\pi)^3} \frac{1}{2\omega_k} \frac{1}{2\pi i} e^{ik(x_1-x_2)} \left[\frac{1}{\omega_k - \omega} + \frac{1}{\omega_k + \omega} \right] \times \sum_\lambda [\varepsilon_\lambda^*(\mathbf{k})]_\mu [\varepsilon_\lambda(\mathbf{k})]_\nu \tag{7.120}$$

The expression in square brackets equals

$$\frac{1}{\omega_k - \omega} + \frac{1}{\omega_k + \omega} = \frac{2\omega_k}{\omega_k^2 - \omega^2} = \frac{2\omega_k}{k^2} \tag{7.121}$$

where $k^2 = \omega_k^2 - \omega^2 = \mathbf{k}^2 - \omega^2$ is the square of the four-momentum of the photon. Substitution of Equation (7.121) in Equation (7.120) gives

$$\mathcal{T}_{\mu\nu}(x_1, x_2) = \frac{1}{i} \int \frac{d^4k}{(2\pi)^4} \frac{e^{ik(x_1-x_2)}}{k^2} \sum_\lambda [\varepsilon_\lambda^*(\mathbf{k})]_\mu [\varepsilon_\lambda(\mathbf{k})]_\nu \tag{7.122}$$

It is seen that the time-ordered product is a function of $x_1 - x_2$ and not of x_1 and x_2 separately. The function is also even as can be shown by replacing $k \to -k$ and following the procedure used above.

The sum over polarizations can be done in analogy with the sum over polarizations for a massive spin 1 field, discussed in Section 6.2. When the photon is off the mass shell, as is the case here, the arguments leading to the exclusion of $\lambda = 3$ in the summation over λ are invalid and the sum is over $\lambda = 1, 2, 3$. We chose three mutually perpendicular polarization vectors as in Equation (6.85). Requiring the Lorenz condition $\partial_\mu A_\mu = 0$ gives $a|\mathbf{k}| - b\omega = 0$. The requirement that $|\varepsilon_3| = 1$ gives that $a^2 - b^2 = 1$. The solution is $a = \omega/\sqrt{-k^2}$ and $b = |\mathbf{k}|/\sqrt{-k^2}$ so $\varepsilon_3 = (0, 0, \omega/\sqrt{-k^2}, i|\mathbf{k}|/\sqrt{-k^2})$, compare with Equation (6.86).

Following the procedure near Equation (6.88) we find

$$\sum_\lambda [\varepsilon_\lambda^*(\mathbf{k})]_\mu [\varepsilon_\lambda(\mathbf{k})]_\nu = \delta_{\mu\nu} - \frac{k_\mu k_\nu}{k^2} \tag{7.123}$$

Note the minus sign in comparison with Equation (6.87). The minus sign is due to the fact that $k^2 = -m^2$. Substitution of Equation (7.123) in Equation (7.122) gives

$$\mathcal{T}_{\mu\nu}(|x_1 - x_2|) = \frac{1}{i} \int \frac{d^4k}{(2\pi)^4} \frac{e^{ik(x_1-x_2)}}{k^2} \left(\delta_{\mu\nu} - \frac{k_\mu k_\nu}{k^2} \right) \tag{7.124}$$

The exponential in Equation (7.124) together with the exponential left out of the amplitudes in Equations (7.107), (7.108), (7.110) and (7.111) give energy-momentum conservation at each vertex and therefore conservation of energy-momentum for the entire process. For example, the exponential in Equation (7.107) together with the exponential in Equation (7.124) gives

$$\int d^4x_1\, d^4x_2\, e^{i(p_1x_1+p_2x_1-p'_2x_2-p'_1x_2)}\, e^{ik(x_2-x_1)} \tag{7.125}$$

where we replaced $x_1 - x_2$ by $x_2 - x_1$ using the fact that $\mathcal{T}$ is even in $x_1 - x_2$, see the remark below Equation (7.122). This is equivalent to replacing k by $-k$. The integrations over x_1 and x_2 in Equation (7.125) give two δ-functions and we obtain

$$\int d^4x_1\, d^4x_2\, e^{i(p_1x_1+p_2x_1-p'_2x_2-p'_1x_2)} e^{ik(x_2-x_1)}$$
$$= (2\pi)^4\, \delta^4(p_1+p_2-k)(2\pi)^4\, \delta^4(-p'_1-p_2+k) \tag{7.126}$$

We conclude that the amplitude is zero unless $k = p_1 + p_2$ *and* $k = p'_1 + p'_2$ so that automatically $p_1 + p_2 = p'_1 + p'_2$. The integration over d^4k in Equation (7.124) is trivial owing to the δ-functions. Similar conclusions are reached for other diagrams. For this reason we do not carry the exponentials in our calculations and they do not appear in Feynman graphs. One must take care to insert the factor $(2\pi)^8$ in the calculation of the transition probability; compare with a factor $(2\pi)^4$ for a first-order calculation, see Equation (7.51).

The term proportional to $k_\mu k_\nu$ in Equation (7.124) will not contribute. An example is provided by the amplitude in Equation (7.107) for $e^+ e^- \to \mu^+\mu^-$. Taking its first factor $\bar{v}_r(\mathbf{p}_1)(e\gamma_\mu)\, u_s(\mathbf{p}_2)$ multiplied by k_μ (with the summation over μ implied) gives (leaving out the factor e)

$$\bar{v}_r(\mathbf{p}_1)\gamma_\mu\, u_s(\mathbf{p}_2)\, k_\mu \tag{7.127}$$
$$= \bar{v}_r(\mathbf{p}_1)\gamma_\mu\, u_s(\mathbf{p}_2)\, [(p_1)_\mu + (p_2)_\mu] \tag{7.128}$$
$$= \bar{v}_r(\mathbf{p}_1)\gamma_\mu(p_1)_\mu\, u_s(\mathbf{p}_2) + \bar{v}_r(\mathbf{p}_1)\gamma_\mu\, (p_2)_\mu u_s(\mathbf{p}_2) \tag{7.129}$$

The spinors satisfy

$$(i\gamma p + m)\, u(\mathbf{p}) = 0 \qquad \bar{u}(\mathbf{p})\, (i\gamma p + m) = 0 \tag{7.130}$$
$$(i\gamma p - m)\, v(\mathbf{p}) = 0 \qquad \bar{v}(\mathbf{p})\, (i\gamma p - m) = 0 \tag{7.131}$$

see Equations (4.108), (4.109), (4.110) and (4.111). Using these relations, one can show that

$$\bar{v}_r(\mathbf{p}_1)\gamma_\mu(p_1)_\mu = \bar{v}_r(\mathbf{p}_1)\frac{m}{i} \qquad \gamma_\mu(p_2)_\mu u_s(\mathbf{p}_2) = -\frac{m}{i}u_s(\mathbf{p}_2) \tag{7.132}$$

Substituting these relations in Equation (7.127) we find that the two terms cancel exactly (even if $m \neq 0$).

Another example is provided by the amplitude in Equation (7.110) for $e^- \mu^+ \to e^- \mu^+$. Taking its first factor $\overline{u}_{r'}(\mathbf{p}'_1)(-e\gamma_\mu)u_r(\mathbf{p}_1)$ multiplied by k_μ (with the summation over ν implied) gives (leaving out the factor $-e$)

$$\overline{u}_{r'}(\mathbf{p}'_1)\gamma_\nu u_r(\mathbf{p}_1)k_\nu \tag{7.133}$$

$$= \overline{u}_{r'}(\mathbf{p}'_1)\gamma_\nu u_r(\mathbf{p}_1)\left[(p_1)_\nu - (p'_1)_\nu\right] \tag{7.134}$$

$$= \overline{u}_{r'}(\mathbf{p}'_1)\gamma_\nu(p_1)_\nu u_r(\mathbf{p}_1) - \overline{u}_{r'}(\mathbf{p}'_1)\gamma_\nu(p'_1)_\nu u_r(\mathbf{p}_1) \tag{7.135}$$

One can show that

$$\overline{u}_{r'}(\mathbf{p}'_1)\gamma_\nu(p'_1)_\nu = -\overline{u}_{r'}(\mathbf{p}'_1)\frac{m}{i} \qquad \gamma_\nu(p_1)_\nu u_r(\mathbf{p}_1) = -\frac{m}{i}u_r(\mathbf{p}_1) \tag{7.136}$$

Substituting these relations in Equation (7.131) we find that the two terms cancel exactly. The cancelation occurs for all vertices where the particle flavor is unchanged, as it is for the electromagnetic interaction and the neutral weak interaction. In these cases, the time-ordered product Equation (7.124) becomes

$$\langle 0|[T[A_\mu(x_1)A_\nu(x_2)]|0\rangle = \frac{1}{i}\int \frac{d^4q}{(2\pi)^4}\, e^{iq(x_2-x_1)}\,\frac{\delta_{\mu\nu}}{q^2} \tag{7.137}$$

where we have used the fact that the time-ordered product is even in $x_1 - x_2$ and we replaced the momentum k by q. The factor $1/q^2$ in Equation (7.133) is sometimes called a propagator as well; it describes the propagation of the photon from x_1 to x_2. The photon indeed propagates the electromagnetic interaction. Compare with Section 2.2 where the Green's function $G(\mathbf{x}_1, \mathbf{x}_2)$ plays the role of propagating the interaction from position $\mathbf{x}_1$ to position $\mathbf{x}_2$; see, for example, Equation (2.29).

We have derived the time-ordered product for the electromagnetic field A_μ. The time-ordered products for W_μ and Z_μ can be derived in nearly the same manner. The difference is that in the plane wave expansions for W_μ in Equation (6.89), for $W_\mu^\dagger$ in Equation (6.90), and for Z_μ in Equation (6.91) we have factors $1/\sqrt{2E_k}$ instead of the factor $1/\sqrt{2\omega_k}$ in the plane wave expansion for A_μ in Equation (1.30). Here $E_k = \sqrt{\mathbf{k}^2 + m^2}$, see the definition of E_p below Equation (6.91). Also the quantity ω (the energy of the electromagnetic field) becomes E (the energy of the W or Z field). Therefore, Equation (7.121) becomes in the case of W_μ or Z_μ

$$\frac{1}{E_k - E} + \frac{1}{E_k + E} = \frac{2E_k}{E_k^2 - E^2} = \frac{2E_k}{k^2 + m^2} \tag{7.138}$$

where $k^2 = \mathbf{k}^2 - E^2$ is the square of the four-momentum of the W or Z. The term in Equation (7.124) proportional to $k_\mu k_\nu$ cancels in the neutral weak interaction because the particle flavor does not change at a vertex. Thus for the neutral weak interaction we find

$$\langle 0|[T[Z_\mu(x_1)Z_\nu(x_2)]|0\rangle = \frac{1}{i}\int \frac{d^4q}{(2\pi)^4}\, e^{iq(x_2-x_1)}\, \frac{\delta_{\mu\nu}}{q^2+m_Z^2} \tag{7.139}$$

The factor $1/(q^2+m_Z^2)$ is sometimes called the Z-propagator as well; it describes the propagation of the Z-boson from x_1 to x_2. The Z-boson indeed propagates the neutral weak interaction.

The cancelation does *not* occur for the charged weak interaction because there the particle flavor changes at a vertex, so their masses are not equal. The $k_\mu k_\nu$ term is proportional to the particle masses, so in the relativistic limit where particle masses can be neglected, the $k_\mu k_\nu$ term in Equation (7.124) can be neglected. In that case the time-ordered product $\langle 0|[T[W_\mu(x_1)W_\nu(x_2)]|0\rangle$ is given by

$$\langle 0|[T[W_\mu(x_1)W_\nu(x_2)]|0\rangle = \frac{1}{i}\int \frac{d^4q}{(2\pi)^4}\, e^{iq(x_2-x_1)}\, \frac{\delta_{\mu\nu}}{q^2+m_W^2} \tag{7.140}$$

The factor $1/(q^2+m_W^2)$ is sometimes called the W-propagator as well; it describes the propagation of the W-boson from x_1 to x_2. The W indeed propagates the charged weak interaction.

Finally, the propagator is represented by a wavy line with a dot at each end and a factor $\delta_{\mu\nu}/(q^2+m^2)$ where $m=0$ for the propagator of the electromagnetic interaction, and $m=m_Z$ or $m=m_W$ for the propagator of the neutral and charged weak interaction respectively. This is shown in Figure 7.1.

PROBLEMS

(1) The Z^0 has numerous decay modes.
   (a) Write down the Feynman graphs for Z^0 decay into a lepton anti-lepton pair and into a quark anti-quark pair. Show the coupling constants at the vertices.
   (b) Write down the amplitudes for these.
   (c) Calculate the decay rate for one of these decay modes.

(2) Derive Equations (7.56).

(3) Prove Equations (7.132) and (7.136) and derive the missing equation involving $\gamma_\mu p_\mu v_r(\mathbf{p})$.

(4) The helicity of a particle may or may not be conserved at a vertex.
   (a) Show that the electromagnetic interaction conserves the helicity of the particles at a vertex.
   (b) What is the case for the weak interaction?

8

Quantum Electrodynamics

8.1 ELECTRON-POSITRON ANNIHILATION

In this section we will evaluate the cross-section for electron-positron annihilation into muon pairs

$$e^+ e^- \to \mu^+ \mu^- \tag{8.1}$$

This reaction is important as a prototype for electron-positron annihilation into other final states, whether lepton pairs or hadrons. The latter case can be understood as electron-positron annihilation into quark anti-quark pairs. Because we assume that quarks and anti-quarks satisfy the Dirac equation, also here the final state consisting of muon pairs is a useful prototype.

We have already discussed the amplitude and Feynman graph for this process in Section 7.3. To calculate the cross-section for $e^+ e^- \to \mu^+ \mu^-$ we use Equation (2.7) for the cross-section in terms of w_{fi}, the transition probability per unit time for the system to go from the initial state $|i\rangle = |e^+ e^-\rangle$ to the final state $|f\rangle = |\mu^+ \mu^-\rangle$. We obtain with the relative velocity $v = 2$

$$d\sigma = \frac{1}{2} dw_{fi} \tag{8.2}$$

Because the volume V will drop out as always, we will not carry factors involving V.

The kinematic variables are as defined in Section 7.3: the four-momenta and spin indices are p_1, r and p_2, s for the e^+ and e^- respectively and p_1', r' and p_2', s' for the μ^+ and μ^- respectively, and their masses are m_1 and m_2 for the e^+, e^- and μ^+, μ^- respectively. The four-momenta are $p_1 = (\mathbf{p}_1, E_1), p_2 = (\mathbf{p}_2, E_2), p_1' = (\mathbf{p}_1', E_1')$, and $p_2' = (\mathbf{p}_2', E_2')$.

The transition probability per unit time is given by Equation (7.53) so we have

$$dw_{fi} = |\langle f|S|i\rangle|^2 \frac{d^3p'_1}{(2\pi)^3}\frac{d^3p'_2}{(2\pi)^3} \tag{8.3}$$

Because the time T will drop out, we will not carry factors involving T. We already noted that this is a second-order process involving the third term in Equation (7.45) so $\langle f|S|i\rangle$ in Equation (8.3) is given by

$$\langle \mu^+\mu^-|S|e^+e^-\rangle = \frac{(-ie)^2}{2}\frac{1}{i}\sqrt{\frac{m_1^2m_2^2}{E_1E'_1E_2E'_2}}(2\pi)^4\delta^4(p_1+p_2-p'_1-p'_2)$$
$$\times\, \bar{v}_r(\mathbf{p}_1)\gamma_\mu\, u_s(\mathbf{p}_2)\frac{\delta_{\mu\nu}}{q^2}\,\bar{u}_{s'}(\mathbf{p}'_2)\gamma_\nu v_{r'}(\mathbf{p}'_1) \tag{8.4}$$

The four-vector q is the four-momentum that we encounter as the integration variable in the integral over the propagator, see Equation (7.137). There are four factors $\sqrt{m/E}$ from the plane wave expansions, and also here we dropped factors involving the volume V. The factor $(2\pi)^4$ results from two factors $(2\pi)^4$ in the numerator from the integrations over x_1 and x_2 in the exponentials and one factor $(2\pi)^4$ in the denominator of the propagator in Equation (7.137), which also gives the factor i in the denominator of Equation (8.4). As discussed in Section 7.4 below Equation (7.108), there are two diagrams that contribute equally, canceling the factor 2 in the denominator of Equation (8.4).

We left out the exponentials from the plane wave expansions of $\bar{\psi}$ and ψ and from the propagator, see Equation (7.137). It is easy to verify that they give factors

$$\int d^4x_1\, d^4x_2\, e^{i(p_1+p_2-q)x_1+i(q-p'_2-p'_1)x_2} \tag{8.5}$$

and

$$\int d^4x_1\, d^4x_2\, e^{i(q-p'_2-p'_1)x_1+i(p_1+p_2-q)x_2} \tag{8.6}$$

respectively. The integrals in Equations (8.5) and (8.6) lead as usual to δ-functions as follows

$$\int d^4x_1\, d^4x_2\, e^{i(p_1+p_2-q)x_1+i(q-p'_2x_2-p'_1x_2)}$$
$$= (2\pi)^8\delta^4(p_1+p_2-q)\delta^4(q-p'_1-p'_2) \tag{8.7}$$

and

$$\int d^4x_1\, d^4x_2\, e^{i(q-p'_2-p'_1)x_1+i(p_1+p_2-q)x_2}$$
$$= (2\pi)^8\delta^4(q-p'_1-p'_2)\delta^4(p_1+p_2-q) \tag{8.8}$$

respectively. The propagator contains an integral over q and this integration gives

$$\int d^4q\, \delta^4(p_1 + p_2 - q)\delta^4(q - p_1' - p_2') = \delta(p_1 + p_2 - p_1' - p_2') \tag{8.9}$$

and

$$\int d^4q\, \delta^4(q - p_1' - p_2')\delta^4(p_1 + p_2 - q) = \delta(p_1 + p_2 - p_1' - p_2') \tag{8.10}$$

respectively, with the proviso that

$$q = p_1 + p_2 \qquad\qquad q = p_1' + p_2' \tag{8.11}$$

These relations show that we have energy-momentum conservation at each vertex and for the process as a whole.

We are interested in the cross-section, integrated over all allowed momenta of the particles in the final states and summed over their spin directions. We also want the cross-section to be averaged over the spin directions of the particles in the initial state. Thus we must square the absolute value of $\langle \mu^+\mu^-|S|e^+e^-\rangle$ in Equation (8.4), sum over the spins of the particles in the initial state and the final state, and divide twice by 2 to get the desired average over spin directions.

The $\delta_{\mu\nu}$ factor in Equation (8.4) makes the indices of the two γ matrices equal. When squaring the expression in Equation (8.4), we multiply it by its complex conjugate. We use ν as the summation index in the complex conjugate. The procedure for evaluating the square of the complex number involving spinors and γ matrices follows the procedure of Section 7.2. Therefore, we will do the calculation in an abbreviated form. The expression $\overline{v}_r(\mathbf{p}_1)\gamma_\mu u_s(\mathbf{p}_2)\overline{u}_{s'}(\mathbf{p}_2')\gamma_\mu v_{r'}(\mathbf{p}_1')$ is itself a product of two complex numbers $\overline{v}_r(\mathbf{p}_1)\gamma_\mu u_s(\mathbf{p}_2)$ and $\overline{u}_{s'}(\mathbf{p}_2')\gamma_\mu v_{r'}(\mathbf{p}_1')$, so we need to multiply each of these two factors by their own complex conjugate. The first factor depends upon the spin indices r and s and not upon r' and s'. The reverse is true for the second factor. We group factors that depend upon the same kinematic variables and define the quantities $\mathcal{L}^e(p_1, p_2)$ and $\mathcal{L}^\mu(p_1', p_2')$ as

$$\mathcal{L}^e_{\mu\nu}(p_1, p_2) = \frac{1}{2}\sum_{r,s}[\overline{v}_r(\mathbf{p}_1)\gamma_\mu u_s(\mathbf{p}_2)][\overline{v}_r(\mathbf{p}_1)\gamma_\nu u_s(\mathbf{p}_2)]^* \tag{8.12}$$

$$\mathcal{L}^\mu_{\mu\nu}(p_1', p_2') = \frac{1}{2}\sum_{r',s'}[\overline{u}_{s'}(\mathbf{p}_2')\gamma_\mu v_{r'}(\mathbf{p}_1')][\overline{u}_{s'}(\mathbf{p}_2')\gamma_\nu v_{r'}(\mathbf{p}_1')]^* \tag{8.13}$$

We have included the two factors $\frac{1}{2}$ from the averaging over the spins of the particles in the initial state in each of the definitions. To get the square of $|\langle \mu^+\mu^-|S|e^+e^-\rangle|$ we must evaluate $\mathcal{L}^e_{\mu\nu}\mathcal{L}^\mu_{\mu\nu}$, a contraction over μ and ν.

The complex conjugation of a complex number that is itself a product of spinors and γ matrices was discussed below Equation (7.54). Using the same procedure, we get for $\mathcal{L}^e(p_1, p_2)$

$$\mathcal{L}^e_{\mu\nu}(p_1, p_2) = \frac{1}{2}\sum_{r,s}[\overline{v}_r(\mathbf{p}_1)\gamma_\mu\, u_s(\mathbf{p}_2)][\overline{v}_r(\mathbf{p}_1)\gamma_\nu\, u_s(\mathbf{p}_2)]^* \tag{8.14}$$

$$= \frac{1}{2}\sum_{r,s}[\overline{v}_r(\mathbf{p}_1)\gamma_\mu\, u_s(\mathbf{p}_2)][v_r^\dagger(\mathbf{p}_1)\gamma_4\gamma_\nu\, u_s(\mathbf{p}_2)]^* \tag{8.15}$$

$$= \frac{1}{2}\sum_{r,s}[\overline{v}_r(\mathbf{p}_1)\gamma_\mu\, u_s(\mathbf{p}_2)][u_s^\dagger(\mathbf{p}_2)\gamma_\nu\gamma_4\, v_r(\mathbf{p}_1)] \tag{8.16}$$

$$= \frac{1}{2}\sum_{r,s}[\overline{v}_r(\mathbf{p}_1)\gamma_\mu\, u_s(\mathbf{p}_2)][\overline{u}_s(\mathbf{p}_2)\gamma_4\gamma_\nu\gamma_4 v_r(\mathbf{p}_1)] \tag{8.17}$$

$$= \frac{1}{2}\sum_{r}\overline{v}_r(\mathbf{p}_1)\gamma_\mu\, \Lambda^+(\mathbf{p}_2)\gamma_4\gamma_\nu\, \gamma_4 v_r(\mathbf{p}_1) \tag{8.18}$$

$$= \frac{1}{2}\sum_{r}\sum_{\alpha\beta}[\overline{v}_r(\mathbf{p}_1)]_\alpha[\gamma_\mu\, \Lambda^+(\mathbf{p}_2)\gamma_4\gamma_\nu\, \gamma_4]_{\alpha\beta}[v_r(\mathbf{p}_1)]_\beta \tag{8.19}$$

$$= \frac{1}{2}\sum_{r}\sum_{\alpha\beta}[v_r(\mathbf{p}_1)]_\beta\, [\overline{v}_r(\mathbf{p}_1)]_\alpha[\gamma_\mu\, \Lambda^+(\mathbf{p}_2)\gamma_4\gamma_\nu\, \gamma_4]_{\alpha\beta} \tag{8.20}$$

$$= -\frac{1}{2}\sum_{\alpha,\beta}[\Lambda^-(\mathbf{p}_1)]_{\beta\alpha}[\gamma_\mu\, \Lambda^+(\mathbf{p}_2)\gamma_4\gamma_\nu\, \gamma_4]_{\alpha\beta} \tag{8.21}$$

$$= -\frac{1}{2}\mathrm{Tr}[\Lambda^-(\mathbf{p}_1)\, \gamma_\mu\, \Lambda^+(\mathbf{p}_2)\gamma_4\gamma_\nu\gamma_4] \tag{8.22}$$

Similarly, we find for $\mathcal{L}^\mu(p_1, p_2)$

$$\mathcal{L}^\mu_{\mu\nu}(p_1', p_2') = \frac{1}{2}\sum_{r,s}[\overline{u}_s'(\mathbf{p}_2')\gamma_\mu\, v_r'(\mathbf{p}_1')][\overline{u}_s'(\mathbf{p}_2')\gamma_\nu\, v_r'(\mathbf{p}_1')]^* \tag{8.23}$$

$$= -\frac{1}{2}\mathrm{Tr}[\Lambda^+(\mathbf{p}_2')\, \gamma_\mu\, \Lambda^-(\mathbf{p}_1')\gamma_4\gamma_\nu\gamma_4] \tag{8.24}$$

We simplify $\gamma_4\gamma_\nu\gamma_4$ by commuting one of the γ_4 with its neighboring γ_ν so that it is adjacent to the other γ_4. Because $\gamma_4\gamma_4 = 1$, their product vanishes. If $\nu = 1, 2, 3$ this would give a minus sign, while if $\nu = 4$ we would get a plus sign. Either combination of these signs will appear simultaneously in $\mathcal{L}^e_{\mu\nu}$ and $\mathcal{L}^\mu_{\mu\nu}$ because they share the $\gamma_4\gamma_\nu\gamma_4$ factor. The net result of the commutation of a γ_4 with its neighboring γ_ν in $\mathcal{L}^e_{\mu\nu}$ and $\mathcal{L}^\mu_{\mu\nu}$ is either two minus signs that cancel each other when $\nu = 1, 2, 3$, or two plus signs when

$\nu = 4$. Thus we obtain

$$\mathcal{L}^{e}_{\mu\nu}(p_1, p_2) = -\frac{1}{2}\mathrm{Tr}[\Lambda^{-}(\mathbf{p}_1)\,\gamma_\mu\,\Lambda^{+}(\mathbf{p}_2)\,\gamma_\nu] \tag{8.25}$$

$$\mathcal{L}^{\mu}_{\mu\nu}(p'_1, p'_2) = -\frac{1}{2}\mathrm{Tr}[\Lambda^{+}(\mathbf{p}'_2)\,\gamma_\mu\,\Lambda^{-}(\mathbf{p}'_1)\,\gamma_\nu] \tag{8.26}$$

To evaluate the traces we simplify the calculation by neglecting the mass terms in the numerators of the expressions Equation (7.57) for Λ^+ and Λ^-. The expressions in Equation (8.25) and Equation (8.26) become

$$\mathcal{L}^{e}_{\mu\nu}(p_1, p_2) = -\frac{1}{8m_1^2}\mathrm{Tr}[\not{p}_1\,\gamma_\mu\,\not{p}_2\,\gamma_\nu] \tag{8.27}$$

$$\mathcal{L}^{\mu}_{\mu\nu}(p'_1, p'_2) = -\frac{1}{8m_2^2}\mathrm{Tr}[\not{p}'_2\,\gamma_\mu\,\not{p}'_1\,\gamma_\nu] \tag{8.28}$$

We have encountered Traces over four γ matrices in Equation (7.72) where we evaluated $\mathrm{Tr}[\not{p}_2\gamma_\mu\,\not{p}_1\gamma_\nu]$ in Equation (7.73) through Equation (7.75), and noted that the result is symmetric under exchange of μ and ν as well as under exchange of p_1 and p_2. In the present case we get, ordering the momenta to our liking

$$\mathcal{L}^{e}_{\mu\nu}(p_1, p_2) = -\frac{1}{2m_1^2}[p_{1\mu}p_{2\nu} + p_{1\nu}p_{2\mu} - (p_1p_2)\delta_{\mu\nu}] \tag{8.29}$$

$$\mathcal{L}^{\mu}_{\mu\nu}(p'_1, p'_2) = -\frac{1}{2m_2^2}[p'_{1\mu}p'_{2\nu} + p'_{1\nu}p'_{2\mu} - (p'_1p'_2)\delta_{\mu\nu}] \tag{8.30}$$

We now evaluate the contraction (sum over μ and ν) of the $\mathcal{L}^{e}_{\mu\nu}$ in Equation (8.29) with $\mathcal{L}^{\mu}_{\mu\nu}$ in Equation (8.30) and obtain

$$\mathcal{L}^{e}_{\mu\nu}(p_1, p_2)\mathcal{L}^{\mu}_{\mu\nu}(p'_1, p'_2) = \frac{1}{2m_1^2m_2^2}[(p_1p'_1)(p_2p'_2) + (p_1p'_2)(p'_1p_2)] \tag{8.31}$$

where we used that $\delta_{\mu\nu}\delta_{\mu\nu} = 4$. The result in Equation (8.31) is seen to be symmetric under exchange of p_1 and p_2 as well as under exchange of p'_1 and p'_2.

The square of the absolute value of Equation (8.4) involves the square of the four-dimensional δ-function which equals $\delta^4/(2\pi)^4$ according to Equation (7.54). We are finally ready to evaluate Equation (8.3) and substitute the result in Equation (8.2) to get

$$d\sigma = \alpha^2\frac{(p_1p'_1)(p_2p'_2) + (p_1p'_2)(p'_1p_2)}{E_1E'_1E_2E'_2\,q^4}\,\delta^4(p_1 + p_2 - p'_1 - p'_2)\,d^3\mathbf{p}'_1d^3\mathbf{p}'_2 \tag{8.32}$$

where we have set $e^2 = 4\pi\alpha$ where α is the fine-structure constant and $q = p_1 + p_2$. In the center-of-mass of the $e^+ e^-$ with θ^* the angle between $\mathbf{p}_1$ and $\mathbf{p}'_1$ (and thus between $\mathbf{p}_2$ and $\mathbf{p}'_2$) we have

$$(p_1 p'_1)(p_2 p'_2) = E_1 E'_1 E_2 E'_2 (\cos\theta^* - 1)^2 \tag{8.33}$$

$$(p_1 p'_2)(p'_1 p_2) = E_1 E'_1 E_2 E'_2 (\cos\theta^* + 1)^2 \tag{8.34}$$

$$q^4 = (p_1 + p_2)^4 = (E_1 + E_2)^4 = s^2 \tag{8.35}$$

where $s = -(p_1 + p_2)^2 = (E_1 + E_2)^2$ is the center-of-mass energy squared and Equation (8.32) becomes

$$d\sigma = 2\alpha^2 \frac{\cos^2\theta^* + 1}{s^2} \delta^4(p_1 + p_2 - p'_1 - p'_2)\, d^3\mathbf{p}'_1 d^3\mathbf{p}'_2 \tag{8.36}$$

We integrate over $\mathbf{p}'_2$ using three of the four δ functions. Using the fact that $\mathbf{p}_1 + \mathbf{p}_2 = 0$ in the $e^+ e^-$ center-of-mass, the δ-function becomes $\delta^3(-\mathbf{p}'_1 - \mathbf{p}'_2)$. As expected, the three-momenta of the particles in the final state are also equal in magnitude and opposite in direction. We neglected particle masses so $E'_1 = E'_2$ and likewise $E_1 = E_2$. We set $E_1 = E_2 = E$. The remaining δ-function reads $\delta(E_1 + E_2 - E'_1 - E'_2) = \delta(2E - 2E'_1) = \frac{1}{2}\delta(E'_1 - E)$. We write $d^3\mathbf{p}'_1 = E'^2_1 dE'_1\, d\cos\theta^*\, d\phi$. The integral over E'_1 is trivial and we get $E'_1 = E$. So all four particles have the same energy, as expected in the limit where masses are neglected.

Because Equation (8.36) does not depend upon ϕ, the integral over ϕ gives 2π. We obtain

$$\frac{d\sigma}{d\cos\theta^*} = \frac{\pi\alpha^2}{2}\frac{1 + \cos^2\theta^*}{s} \tag{8.37}$$

where we used that the center-of mass energy squared $s = (2E)^2$. Integrating Equation (8.37) over $\cos\theta^*$ we get

$$\sigma = \frac{4\pi\alpha^2}{3s} \tag{8.38}$$

for the total cross-section. Reinserting $\hbar$ and c we obtain

$$\sigma = \frac{4\pi\alpha^2(\hbar c)^2}{3s} \tag{8.39}$$

We could have guessed the α^2 and s dependence from the onset of the calculation by counting vertices and dimensional analysis respectively. The cross-section for other final state particles can be obtained immediately from Equation (8.39) by a suitable substitution of coupling constant(s).

8.2 ELECTRON-MUON SCATTERING

In this section we will evaluate the cross-section for elastic electron-muon scattering. For definiteness we consider

$$e^- \mu^+ \to e^- \mu^+ \tag{8.40}$$

This reaction is important as a prototype for elastic and inelastic lepton-nucleon scattering where the lepton is either an electron or a muon of either charge. Inelastic lepton-nucleon scattering can be understood as an incoherent sum of elastic lepton-quark and lepton-antiquark scattering. Because we assume that quarks and anti-quarks satisfy the Dirac equation, the reaction in Equation (8.40) is useful for an understanding of inelastic lepton-nucleon scattering. The calculation of the reaction in Equation (8.40) follows closely the calculation of $e^+ e^- \to \mu^+ \mu^-$ in the previous section, so we will make use of some of those results below.

We have already discussed the amplitude and Feynman graph for this process in Section 7.3. The kinematic variables are as defined in Section 7.3: the four-momenta and spin indices are p_1, r and p_2, s for the e^- and μ^+ in the initial state respectively and p_1', r' and p_2', s' for the e^- and μ^+ in the final state respectively, and their masses are m_1 and m_2 for the e^- and μ^+ respectively. The four-momenta are $p_1 = (\mathbf{p}_1, E_1)$, $p_2 = (\mathbf{p}_2, E_2)$, $p_1' = (\mathbf{p}_1', E_1')$, and $p_2' = (\mathbf{p}_2', E_2')$.

To calculate the cross-section for $e^- \mu^+ \to e^- \mu^+$ we use Equation (8.2) for the cross-section in terms of w_{fi} and Equation (8.3) for the dw_{fi}, the transition probability per unit time for the system to go from the initial state $|i\rangle = |e^- \mu^+\rangle$ to the final state $|f\rangle = |e^- \mu^+\rangle$. We obtain

$$d\sigma = \frac{1}{2} |\langle f|S|i\rangle|^2 \frac{d^3 p_1'}{(2\pi)^3} \frac{d^3 p_2'}{(2\pi)^3} \tag{8.41}$$

with $\langle f|S|i\rangle$ given by Equation (7.110)

$$\langle e^- \mu^+ |S| e^- \mu^+\rangle = \frac{(-ie)^2}{2} \frac{1}{i} \sqrt{\frac{m_1^2 m_2^2}{E_1 E_1' E_2 E_2'}} (2\pi)^4 \delta^4(p_1 + p_2 - p_1' - p_2')$$
$$\times \bar{u}_{r'}(\mathbf{p}_1') \gamma_\mu u_r(\mathbf{p}_1) \frac{\delta_{\mu\nu}}{q^2} \bar{v}_s(\mathbf{p}_2) \gamma_\nu v_{s'}(\mathbf{p}_2') \tag{8.42}$$

As discussed in Section 7.4 there are two diagrams that contribute equally, canceling the factor 2 in the denominator of Equation (8.42). In the present case

$$q = p_1 - p_1' \qquad \text{and} \qquad q = p_2' - p_2 \tag{8.43}$$

as can be seen from the exponentials that appear from the plane wave expansions and the propagator; compare with Equation (8.11). These relations show that we have energy-momentum conservation at each vertex and for the process as a whole.

We are interested in the cross-section, differential in q^2 and integrated over all allowed momenta of the particles in the final states and summed over their spin directions. We also want the cross-section to be averaged over the spin directions of the particles in the initial state. Thus we must square the absolute value of $\langle\mu^+\mu^-|S|e^+e^-\rangle$ in Equation (8.42), sum over the spins of the particles in the initial state and the final state, and divide twice by 2 to get the desired average over spin directions.

As before, the $\delta_{\mu\nu}$ factor in Equation (8.42) makes the indices of the two γ matrices equal. When squaring the expression in Equation (8.42), we proceed as before and introduce $\mathcal{L}^e(p_1, p_1')$ and $\mathcal{L}^\mu(p_2, p_2')$ as

$$\mathcal{L}^e_{\mu\nu}(p_1, p_1') = \frac{1}{2}\sum_{r,r'}[\overline{u}_{r'}(\mathbf{p}_1')\gamma_\mu\, u_r(\mathbf{p}_1)][\overline{u}_{r'}(\mathbf{p}_1')\gamma_\nu\, u_r(\mathbf{p}_1)]^* \tag{8.44}$$

$$\mathcal{L}^\mu_{\mu\nu}(p_2, p_2') = \frac{1}{2}\sum_{s,s'}[\overline{v}_s(\mathbf{p}_2)\gamma_\mu v_{s'}(\mathbf{p}_2')][\overline{v}_s(\mathbf{p}_2)\gamma_\nu v_{s'}(\mathbf{p}_2')]^* \tag{8.45}$$

We have included the two factors $\frac{1}{2}$ from the averaging over the spins of the particles in the initial state in each of the definitions. To get the square of $|\langle\mu^+\mu^-|S|e^+e^-\rangle|$ we must evaluate $\mathcal{L}^e_{\mu\nu}\mathcal{L}^\mu_{\mu\nu}$. We obtain

$$\mathcal{L}^e_{\mu\nu}(p_1, p_1') = \frac{1}{2}\mathrm{Tr}[\Lambda^+(\mathbf{p}_1')\,\gamma_\mu\,\Lambda^+(\mathbf{p}_1)\gamma_4\gamma_\nu\gamma_4] \tag{8.46}$$

$$= \frac{1}{2}\mathrm{Tr}[\Lambda^+(\mathbf{p}_1')\,\gamma_\mu\,\Lambda^+(\mathbf{p}_1)\gamma_\nu] \tag{8.47}$$

and

$$\mathcal{L}^\mu_{\mu\nu}(p_2, p_2') = \frac{1}{2}\mathrm{Tr}[\Lambda^-(\mathbf{p}_2)\,\gamma_\mu\,\Lambda^-(\mathbf{p}_2')\gamma_4\gamma_\nu\gamma_4] \tag{8.48}$$

$$= \frac{1}{2}\mathrm{Tr}[\Lambda^-(\mathbf{p}_2)\,\gamma_\mu\,\Lambda^-(\mathbf{p}_2')\gamma_\nu] \tag{8.49}$$

As we must calculate $\mathcal{L}^e_{\mu\nu}\mathcal{L}^\mu_{\mu\nu}$ we commuted a γ_4 with the γ_ν simultaneously in Equation (8.46) and Equation (8.48), see the discussion below Equation (8.24).

As before, we neglect the mass terms in the numerators of the expressions Equation (7.57) for Λ^+ and Λ^-. The expressions Equation (8.47) and Equation (8.49) become

$$\mathcal{L}^e_{\mu\nu}(p_1, p_2) = -\frac{1}{8m_1^2}\mathrm{Tr}[\not{p}_1'\,\gamma_\mu\,\not{p}_1\,\gamma_\nu] \tag{8.50}$$

and

$$\mathcal{L}^{\mu}_{\mu\nu}(p_2, p_2') = -\frac{1}{8m_2^2}\mathrm{Tr}[\not{p}_2\,\gamma_\mu\,\not{p}_2'\,\gamma_\nu] \tag{8.51}$$

We have encountered Traces over four γ matrices in Equation (7.72) and in Equation (8.27) and Equation (8.28). We can use any of these results with a proper substitution of variables to get, ordering the momenta to our liking,

$$\mathcal{L}^{e}_{\mu\nu}(p_1, p_1') = -\frac{1}{2m_1^2}[p_{1\mu}p'_{1\nu} + p_{1\nu}p'_{1\mu} - (p_1p_1')\delta_{\mu\nu}] \tag{8.52}$$

$$\mathcal{L}^{\mu}_{\mu\nu}(p_2, p_2') = -\frac{1}{2m_2^2}[p_{2\mu}p'_{2\nu} + p_{2\nu}p'_{2\mu} - (p_2p_2')\delta_{\mu\nu}] \tag{8.53}$$

We now evaluate the contraction (sum over μ and ν) of $\mathcal{L}^{e}_{\mu\nu}$ in Equation (8.52) with $\mathcal{L}^{\mu}_{\mu\nu}$ in Equation (8.53) and obtain directly from Equation (8.31) with a proper substitution of variables

$$\mathcal{L}^{e}_{\mu\nu}(p_1, p_1')\mathcal{L}^{\mu}_{\mu\nu}(p_2, p_2') = \frac{1}{2m_1^2m_2^2}[(p_1p_2)(p_1'p_2') + (p_1p_2')(p_1'p_2)] \tag{8.54}$$

The result in Equation (8.54) is seen to be symmetric under exchange of p_1 and p_1' as well as under exchange of p_2 and p_2'.

We are finally ready to evaluate Equation (8.41) and substitute the result in Equation (8.41) to get

$$d\sigma = \alpha^2 \frac{(p_1p_2)(p_1'p_2') + (p_1p_2')(p_1'p_2)}{E_1E_1'E_2E_2'\,q^4}\,\delta^4(p_1 + p_2 - p_1' - p_2')\,d^3\mathbf{p}_1'd^3\mathbf{p}_2' \tag{8.55}$$

where we have set $e^2 = 4\pi\alpha$ where α is the fine-structure constant and q is defined in Equation (8.43). In the center-of-mass of the $e^-\ \mu^+$ with θ^* the angle between $\mathbf{p}_1$ and $\mathbf{p}_1'$ (and thus between $\mathbf{p}_2$ and $\mathbf{p}_2'$) we have

$$(p_1p_2)(p_1'p_2') = 4E_1E_1'E_2E_2' \tag{8.56}$$

$$(p_1p_2')(p_1'p_2) = E_1E_1'E_2E_2'\,(\cos\theta^* + 1)^2 \tag{8.57}$$

$$q^2 = 4E_1E_1'\sin^2\tfrac{1}{2}\theta^* + 2m_1^2 \tag{8.58}$$

where $q = p_1 - p_1'$ is called the momentum transfer (four-momentum is transferred from the electron to the muon), compare with Equation (2.36). Substitution of Equation (8.56) and Equation (8.57) in Equation (8.55) gives

$$d\sigma = \alpha^2\,\frac{4 + (\cos\theta^* + 1)^2}{q^4}\,\delta^4(p_1 + p_2 - p_1' - p_2')\,d^3\mathbf{p}_1'd^3\mathbf{p}_2' \tag{8.59}$$

We integrate over $\mathbf{p}_2'$ using three of the four δ functions and find that $\mathbf{p}_2' = -\mathbf{p}_1'$ and thus that $E_2' = E_1' = E$. We integrate over E_2' using the remaining δ-function $\delta(2E - 2E_1') = -\frac{1}{2}\delta(E_1' - E)$ as in the previous Section below Equation (8.36). We obtain

$$\frac{d\sigma}{d\cos\theta^*} = \pi\alpha^2 E^2 \frac{4 + (\cos\theta^* + 1)^2}{q^4} = 4\pi\alpha^2 E^2 \frac{1 + (\cos\frac{1}{2}\theta^*)^4}{q^4} \tag{8.60}$$

For small scattering angles this expression shows the well known $(1/\theta^*)^4$ dependence of Coulomb scattering that we encountered in Section 2.2, Equation (2.55).

In inelastic lepton-nucleon scattering one introduces the scaling variable y defined as $y = (1 - \cos\theta^*)/2$. We also introduce the center-of-mass energy squared $s = (2E)^2$. In terms of the variables y and s we get

$$\frac{d\sigma}{dy} = \frac{2\pi\alpha^2 s}{q^4}[1 + (1 - y)^2] \tag{8.61}$$

This result is the starting point for a discussion of 'Scaling' in inelastic lepton-nucleon scattering. Inserting $\hbar$ and c we obtain instead of Equation (8.60)

$$\frac{d\sigma}{d\cos\theta^*} = \pi\alpha^2(\hbar c)^2 E^2 \frac{4 + (\cos\theta^* + 1)^2}{q^4} \tag{8.62}$$

and instead of Equation (8.61)

$$\frac{d\sigma}{dy} = \frac{2\pi\alpha^2(\hbar c)^2 s}{q^4}[1 + (1 - y)^2] \tag{8.63}$$

PROBLEMS

(1) Consider muon-neutrino electron ($\nu_\mu e^-$) scattering.
  (a) Draw the Feynman graph(s) that govern this process.
  (b) Write down the amplitude for this process.
  (c) What changes if you select a $\overline{\nu}_\mu$ instead of a ν_μ as incident particle? What if you replace the electron by a positron? [The latter is not very practical from an experimental point of view.]
  (d) Calculate the cross-section $d\sigma/d\cos\theta^*$ and $d\sigma/dy$ for case (a) where θ^* and y are defined in Section 8.2. Think carefully about the sum over the spin of the neutrino in the final state. What happens to the average over the spin of the incident ν_μ?

(e) What changes if you select a muon with the opposite sign? What if you replace the electron by a positron?
(f) Calculate the total cross-section from the result in (d). Write the result in terms of G_F and s where s is defined in Section 8.2. Comment on the energy dependence of the total cross-section.

(2) In problem (1) the incident particles were chosen to be from different doublets. Now consider $\nu_e e^-$ scattering.
(a) Draw the Feynman graph(s) for this process and compare with Problem (1).
(b) Now consider $\overline{\nu_e} e^-$ scattering and draw the Feynman graph(s). Again, compare with problems (1) and (2a).
(c) What changes if you replace the electron flavor by the muon flavor for both particles in (b)?
(d) Write down the amplitude for the process specified in (b).
(e) Calculate the cross-section $d\sigma/d\cos\theta^*$ and $d\sigma/dy$ for case (a) where θ^* and y are defined in Section 8.2. Think carefully about the sum over the spin of the neutrino in the final state and the average over the spin of the incident neutrino.
(f) Calculate the total cross-section from the result in (e). Write the result in terms of G_F and s where s is defined in Section 8.2.

(3) In our treatment of $e^+ e^- \to \mu^+ \mu^-$ through an intermediate photon we neglected the amplitude with an intermediate Z^0 in place of the intermediate photon.
(a) Draw the two Feynman graphs that govern $e^+ e^- \to \mu^+ \mu^-$ and write down the amplitude.
(b) Calculate $d\sigma/d\cos\theta^*$ and note the angular dependence.
(c) Calculate the total cross-section and note the energy dependence.
(d) Use the results obtained to write down $d\sigma/d\cos\theta^*$ and the total cross-section for $e^+ e^-$ annihilation into a given flavor of quark anti-quark pairs.

(4) Use the definitions $a_R = \frac{1}{2}(1-\gamma_5), a_L = \frac{1}{2}(1+\gamma_5), \overline{\psi} = \psi^\dagger\gamma_4$, and $\psi_L = a_L\psi$, and the properties $\gamma_5\gamma_\mu + \gamma_\mu\gamma_5 = 0$ $\mu = 1,2,3,4$ and $(\gamma_5)^2 = 1$.
(a) Show that $a_R + a_L = 1$.
(b) Show that $a_R a_L = a_L a_R = 0$.
(c) Show that $(a_R)^2 = a_R$ and $(a_L)^2 = a_L$.
(d) Show that the only non-zero expressions of the form $a_R\Gamma_a a_L$ with Γ_a one of the five possibilities $1, \gamma_\mu, \frac{1}{2}i(\gamma_\mu\gamma_\nu - \gamma_\nu\gamma_\mu), \gamma_5\gamma_\mu, \gamma_5$ are $a_R\gamma_\mu a_L = \gamma_\mu a_L$ and $a_R\gamma_\mu\gamma_5 a_L = -\gamma_\mu a_L$.
(e) Show that $\overline{\psi}_L = \overline{\psi}\, a_R$.
(f) Show that $\overline{\psi}_L\Gamma_a\psi_L = \overline{\psi} a_R\Gamma_a a_L\psi$.

Index

An Introduction to Advanced Quantum Physics Hans P. Paar

Printed and bound by CPI Group (UK) Ltd, Croydon, CR0 4YY

27/10/2024

14580160-0002